MU120
Open Mathematics

The Open University

GW00697383

Unit 3

Earnings

MU120 course units were produced by the following team:

Gaynor Arrowsmith (Course Manager)
Mike Crampin (Author)
Margaret Crowe (Course Manager)
Fergus Daly (Academic Editor)
Judith Daniels (Reader)
Chris Dillon (Author)
Judy Ekins (Chair and Author)
John Fauvel (Academic Editor)
Barrie Galpin (Author and Academic Editor)
Alan Graham (Author and Academic Editor)
Linda Hodgkinson (Author)
Gillian Iossif (Author)
Joyce Johnson (Reader)
Eric Love (Academic Editor)
Kevin McConway (Author)
David Pimm (Author and Academic Editor)
Karen Rex (Author)

Other contributions to the text were made by a number of Open University staff and students and others acting as consultants, developmental testers, critical readers and writers of draft material. The course team are extremely grateful for their time and effort.

The course units were put into production by the following:

Course Materials Production Unit (Faculty of Mathematics and Computing)

Martin Brazier (Graphic Designer)
Hannah Brunt (Graphic Designer)
Alison Cadle (TEXOpS Manager)
Jenny Chalmers (Publishing Editor)
Sue Dobson (Graphic Artist)
Roger Lowry (Publishing Editor)

Diane Mole (Graphic Designer)
Kate Richenburg (Publishing Editor)
John A.Taylor (Graphic Artist)
Howie Twiner (Graphic Artist)
Nazlin Vohra (Graphic Designer)
Steve Rycroft (Publishing Editor)

The Open University, Walton Hall, Milton Keynes, MK7 6AA.

First published 1996. Second edition 2005. Reprinted 2006.

Edited, designed and typeset by The Open University, using the Open University TEX System.

Printed and bound in the United Kingdom by The Charlesworth Group, Wakefield.

ISBN 0 7492 0293 9

2.2

Contents

Study guide

This unit consists of six sections. Section 1 involves watching a short piece of video, to raise some of the issues that are investigated in the unit. However, if you are unable to watch the video at the beginning, this need not prevent you from starting Section 2. Sections 2 to 5 are closely linked and should be studied in order. Section 6 contains an audio and may be studied at any time after Section 2. For Section 4 and the Appendix, you will need some graph or squared paper on which to draw boxplots.

As you work through the unit, you will once again be applying mathematical skills from *Unit 2* such as calculating percentages. Averages will be discussed further and the idea of an index will again be used—this time to measure changes in earnings. You should also be concentrating on improving your written communication, and developing and applying the more general features of statistical problem solving, as well as representing and interpreting information.

Remember to make notes of important topics and key ideas as you come across them. A Handbook activity sheet has been provided for you to make notes on new terms, but this sheet represents only a bare minimum of what you should record.

You may find it useful to devise a plan of action for studying this unit, taking into account the different components that it includes and how you are managing your other priorities and commitments alongside your Open University study. Remember to include time for doing the assignment questions within your study time.

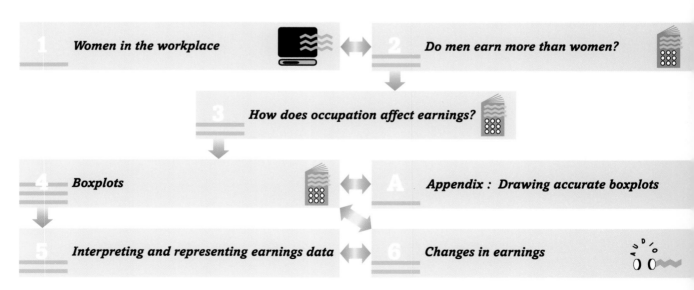

1	Women in the workplace
2	Do men earn more than women?
3	How does occupation affect earnings?
4	Boxplots
A	Appendix : Drawing accurate boxplots
5	Interpreting and representing earnings data
6	Changes in earnings

Summary of sections and other course components needed for *Unit 3*

Introduction

'Are people getting better off?' is the question that was posed at the beginning of *Unit 2*. Two key elements in answering this question were identified there: changes in prices and in earnings. This led to an investigation into how to measure prices and changes in prices. (The unit looked at the position in the UK, but a similar analysis could be used for other countries too.) This unit turns to earnings and begins with a video band about women in the workplace. This raises a number of questions, some of which are investigated in the rest of the unit. For example, 'Do men still earn more than women?' and, if so, 'Has the gap between men's and women's earnings been closing?' Tackling questions such as these is not as straightforward as it might seem. At various points in the text, you will be encouraged to look critically at the data themselves and conclusions drawn from them.

Throughout this unit there are many visual images, representations and diagrams. As you meet them, consider how clear an illustration of an idea they provide, and how they support the main point as you see it.

1 Women in the workplace

Aims   The main aim of this section is to consider how the situation of women who go out to work has changed since the Second World War, and to discuss how statements about women's and men's earnings might be investigated.   ◇

Your work in this section includes a video band about 'Bella the Welder', in which Bella talks of her work during the Second World War (1939–45).

Now watch band 2 of DVD00107 called 'Women in the workplace'.

At the end of this video sequence, you were asked the following questions.

◇   What information have you remembered?

◇   What question would you most like to ask Bella?

The purpose of these questions was to encourage you to be precise about what you noticed and to think about what further information you would like to have. What sorts of question did watching and listening to Bella raise for you? For instance:

◇   Did you want to know more about working conditions for women in the war?

◇   Did you wonder how changes in attitudes to women and work came about in the years following the war?

In the video commentary, a distinction is made between *quantitative* data (information based on measurements, usually given as numbers) and *qualitative* data (information in non-numerical form). Quantitative data are usually more straightforward to record and process. But qualitative data can be altogether more subtle and difficult to interpret. Some information is conveyed both in numbers and words—for example, Bella's earnings and hours worked. Other information is conveyed in her voice and expression. Did you notice how the tone of Bella's voice and her expression altered when she described the completed ship and called it 'a thing of beauty, a wonderful thing'? There is a wealth of information here about her feeling of pride and awe in her achievement and that of her colleagues—information which could not be gained from a written transcript of her words alone.

Traditionally, statistics has focused on information which can be quantified, but that is only a small subset of the wide variety of information that we process when functioning as thinking and feeling human beings. Decisions are regularly made not just on facts that can be written down, but also on how a person looks, speaks, acts, and so on. Your reaction to what Bella had to say was almost certainly influenced by the way she said it.

You have heard how, during the Second World War, male and female welders in the shipyards were paid different rates for doing the same work. Nowadays, this would be against the law in the UK. The Equal Pay Act of 1970, fully implemented in 1975, provides for 'equal pay for men and women doing work of a broadly similar nature'. And the Equal Pay Amendment Regulations (1984) gave women the right to claim 'equal pay for work of equal value to a man's in terms of effort, skill and training'. However, the gap between women's and men's pay is still a matter for discussion despite the legislation.

Activity 1   *The earnings gap*

Think about what data you might need to investigate the extent to which the Equal Pay legislation has eliminated the inequalities between men's and women's pay in Britain. How could you find out how much things have improved since Bella's working days?

To investigate such questions, you would obviously need data on the earnings of men and women. But, in more detail, *which* data?

◇   Is it enough to use data on *average* earnings? If so, what sort of average? Median, mean or something else?

◇   Two people might be paid at the same hourly rate, but one of them may work more hours per week and hence be paid more per week. So the required data might be for hourly earnings, weekly earnings or both.

◇   The legislation refers to equal pay for work 'of equal value'. So separate data will be needed for different types of work.

These issues are addressed in the rest of the unit.

Useful data were published annually from 1970 to 2003 in a UK Government publication called the *New Earnings Survey*. These data could be used to investigate women's and men's earnings during this period.

The Office for National Statistics has replaced the New Earnings Survey by a similar study called the Annual Survey of Hours and Earnings (ASHE). This survey was first carried out in 2004 and covers the whole of the UK. The first publication of ASHE results was not available in time for the revision of this unit, and in any case standard ASHE results do not provide as much detail as is considered here.

The New Earnings Survey

The *New Earnings Survey* was carried out by various Government departments (most recently the Office for National Statistics) each April from 1970 to 2003. Its main purpose was to provide information for the British government about earnings within Great Britain. A similar but separate survey was carried out annually in Northern Ireland. The survey provided information on approximately 160 000 men and women who were almost all members of the Inland Revenue pay-as-you-earn (PAYE) scheme; it did not cover the self-employed, nor, of course, the unemployed.

Each member of the PAYE scheme has a National Insurance number such as YR 04 73 49 D. In any year, all employees whose National Insurance numbers end with a particular pair of digits were included in the survey. (The final *letter* was ignored.) The employer of each selected employee was contacted by post and was required by law to provide certain information. The data collected included the following information concerning the employee.

◇   Gender and age.

◇   Location of workplace.

◇   Occupation.

◇   Total gross weekly earnings during a specified week. (For monthly-paid workers, equivalent weekly earnings were calculated from their month's salary.)

◇   Information about normal basic hours, overtime earnings and hours, and bonus and other incentive payments.

The information collected was analysed by government statisticians and the results are published in a number of volumes, online at www.statistics.gov.uk.

In Sections 2 to 5, data from the 2003 *New Earnings Survey* (Volumes A and D) will be used to investigate the question 'Do men earn more than women?'

Outcomes

After studying this section, you should be able to:

◇   suggest some types of data needed to investigate claims and comparisons between women's and men's earnings (Activity 1).

2 Do men earn more than women?

Aims   The main aim of this section is to show how the mean and the median may be used to make meaningful comparisons between the earnings of men and women.   ◇

In Sections 2 to 5 of this unit, you are going to concentrate on the question 'Do men earn more than women?' There will also be more opportunity for you to work with the statistical problem-solving processes mentioned in *Unit 2*. In this section, the choice of appropriate data on the earnings of men and women and how to compare them is discussed.

2.1   Data—comparing like with like

Differences in earnings between individuals can be accounted for by a great variety of factors apart from gender.

▶   Spend a minute or two thinking about what these factors might be and write down your ideas. (Try to write down at least two ideas before you read on.)

You may have suggested factors such as education, training, social background, age, occupation and number of hours worked. Some people earn more than others for doing similar work. Some jobs are rewarded more than others: danger, level of responsibility and skill required are all factors affecting pay. Personal qualities such as aptitude, ability and temperament also have some effect. You may have thought of others, such as geographical location, nepotism, status, historical precedent: there is a multitude of factors.

Some factors may be more important than gender in determining levels of earnings, some less important. A number of these factors may be interrelated with gender, producing even more complexity. For example:

◇   more women than men work part-time, so some difference between the earnings of men and women may be due to this;

◇   some jobs are done predominantly by women, others mainly by men, and earnings vary historically from job to job, so this is another possible reason for differences in earnings.

Since the various factors which affect earnings are numerous and interrelated, it is impossible to disentangle them. However, the earnings of men and women who are similar with respect to other factors can be compared, and some possible sources of distortion in the results obtained will be eliminated. For example, it would be better to compare men and women working full-time in the same occupation. As far as possible, it is good to *compare like with like*. But, since the investigation depends on published data, which factors *can* be taken into account is to some extent limited. This is an important point to bear in mind if you are planning to carry out your own statistical investigation.

In 2003, more women than men worked part-time. To attempt to eliminate the effect that this factor has on the relative pay of men and women, the investigation will only concern men and women working full-time; and it only applies to employees being paid adult rates whose pay was not affected by absence in the pay-period for which the data were collected. (Over 97% of both men and women surveyed in the *New Earnings Survey* were being paid adult rates.)

All the data in Sections 2 to 5 relate to men and women in Great Britain working full-time on adult pay rates and whose pay was unaffected by absence. The Office for National Statistics did publish some information from the Great Britain and the Northern Ireland *New Earnings Surveys* for the whole of the UK, but they have not published UK-wide data on all the topics discussed in this unit.

Note that identifying the salient variables and deciding on appropriate measures of them is an important aspect of any statistical (and often scientific) investigation.

Before considering the effect of occupation on pay, first consider any difference between the overall earnings of men and women. Table 1 gives the mean gross weekly earnings of adult men and women in full-time employment in Great Britain in 2003.

Table 1   Mean gross weekly earnings of adult men and women in full-time employment (to the nearest pound)

	Women	Men
Mean	396	525

Source: *New Earnings Survey*, 2003, Table A1

Having obtained some data, the next step is to decide how to compare the earnings of men and women. From Table 1, it is clear that the mean gross weekly earnings for men is greater than the corresponding figure for women. Two methods of comparing men's and women's mean earnings probably spring to mind: subtracting one from the other to find the (numerical) difference, and dividing one by the other to find the ratio.

Activity 2   *Comparing men's and women's earnings*

Compare the mean gross weekly earnings of men and women by a method of your choosing. Make a brief note to explain why you chose that method.

First, consider the *absolute* numerical difference: this is £$(525 - 396)$ = £129. So the mean gross weekly earnings of adult men is £129 more than the mean gross weekly earnings for women.

▶   Is this a useful way of comparing the earnings of men and women?

Suppose that the mean weekly earnings of men and women had been £225 and £96, respectively; in this case, the *absolute* numerical difference would also have been £129, as it would have been if the weekly earnings had been £1525 and £1396. However, an absolute difference of £129 would be regarded as of much greater importance in the first case than in the second. So it would be better to know something about the *relative* size of the difference and not just the *absolute* difference. Recall from Chapter 1 of the *Calculator Book*, the discussion of price increases in a pint of milk and a car between 1984 and 1994. Finding the absolute difference there was not too useful a comparison, and the same may be true here. But to talk about the 'gap' between men's and women's earnings may unhelpfully suggest absolute numerical difference as the appropriate measure.

In *Unit 2*, ratios rather than absolute differences were used to compare prices. One of the benefits of using ratios is that whatever the unit of measurement—pounds, pence, euros, dollars—the ratio remains the same. So whatever the unit of measurement used to record earnings, the ratio of women's earnings to men's earnings would be the same. Recall the loaf of bread measure in Section 1 of *Unit 2*: you were able to compare ratios for 1594 and today even though the units of measurement were different—old pence in 1594 and new pence today. Moreover, you were able to do this even though both prices and earnings had changed enormously over the four-hundred-year period. The ratios did not depend on the *absolute* size of the quantities being compared, only on their *relative* size. So ratios calculated at different times can be meaningfully compared. Using relative comparisons makes it possible to extend the investigation to make international comparisons or to make comparisons over time.

So let's use earnings *ratios*. Since the available data are the *mean* gross weekly earnings for men and for women, take the ratio of these means first. The technical term for this is the *earnings ratio at the mean*, and it is defined as follows.

You will use other ratios later.

The *earnings ratio at the mean* is: $\dfrac{\text{mean earnings of women}}{\text{mean earnings of men}}$.

An established convention is to take the earnings ratio at the mean as the mean women's earnings divided by the mean men's earnings, rather than the other way round. Consider an example.

Example 1   *Calculating an earnings ratio at the mean*

For the data in Table 1, the earnings ratio at the mean is

$$\frac{396}{525} = 0.7542857 = 0.75 \text{ (to two decimal places).}$$

In fact, earnings ratios are usually expressed as percentages. Thus, the mean gross weekly earnings of adult women in full-time employment in 2003 was approximately 75% of the mean gross weekly earnings of adult men in full-time employment.

In a context where men usually earn more than women, the earnings ratio at the mean will usually be less than one or, as a percentage, less than 100. The nearer the earnings ratio at the mean is to 100%, the nearer 'average' earnings of women are to those of men.

Table 1 gives the mean gross (that is, before any deductions, such as tax, pension, national insurance, are removed) weekly earnings of adult men and women in full-time employment. This 'compares like with like' to some extent, by avoiding the effects of part-time work and of being paid on non-adult rates. However, there are other factors that affect earnings, like total hours worked, amount of overtime and occupation.

▶   Might any of these factors have an effect on the relative earnings of men and women?

If they do, then, in order to make a fair comparison, they should be taken into account. How can you find out what effect they have, and take the effects into account by excluding them from the comparison?

Look first at hours worked and overtime. You might expect there could be differences between the hours worked and overtime of men and women.

Table 2   Mean weekly hours worked by adult men and women in full-time employment in Great Britain in 2003 (to one decimal place)

	Women	Men
Normal basic	36.7	38.7
Overtime	0.7	2.2
Total	37.4	40.9

Source: *New Earnings Survey*, 2003, Table A1

Activity 3   *Hours and overtime*

In this context, the everyday phrase 'on average' means using the mean.

(a)   On average, how many hours did women work per week in 2003? Was this more or less than the average number of hours worked by men?

(b)   On average, did men or women do more overtime per week in 2003, and by how much?

(c)   What do you think the effect would be of excluding overtime pay from the mean gross weekly earnings used to calculate the earnings ratio at the mean? Do you think this earnings ratio would increase or decrease if overtime were excluded?

Comments on Activities begin on page 65.

(d)   Men and women worked a different number of hours per week on average. Can you suggest a way of eliminating any effect due to this?

Table 3 below gives data excluding overtime for Great Britain in 2003.

Table 3   Mean gross weekly and hourly earnings excluding overtime of adult men and women

	Women	Men
Mean gross weekly earnings excluding overtime (£)	389	500
Mean gross hourly earnings excluding overtime (pence)	1056	1288

Source: *New Earnings Survey*, 2003, Tables A37 and A39.

Activity 4   *Calculating earnings ratios*

In Example 1, the earnings ratio at the mean based on gross weekly earnings including overtime was found to be 75%. Use the data in Table 3 to:

(a)   calculate the earnings ratio at the mean based on gross weekly earnings excluding overtime;

(b)   calculate the earnings ratio at the mean based on gross *hourly* earnings excluding overtime;

(c)   describe the effect on the earnings ratio at the mean of excluding overtime and using data for hourly earnings instead of weekly earnings.

Removing overtime pay from gross weekly earnings increases the earnings ratio at the mean from 75% to 78%. Comparing hourly earnings instead of weekly earnings increases the earnings ratio at the mean further to 82%. So the longer average working week and extra overtime worked by men account for part of the difference between the weekly earnings of men and women. Therefore, to find out whether or not groups of men and women receive equal pay for a similar amount of work, use gross hourly earnings excluding overtime (when available) for a fairer comparison.

2.2   *Has the 'gap' between men's and women's earnings been closing?*

To investigate this question, data on earnings are needed for a number of different years. Table 4 (overleaf) shows the mean gross hourly earnings excluding overtime for adult employees for various years between 1989 and 2003.

Table 4   Mean gross hourly earnings excluding overtime (in pence)

Year	Women	Men
1989	478	629
1991	589	757
1993	669	847
1995	716	901
1996	750	939
1997	788	982
1998	824	1031
1999	870	1075
2000	913	1126
2001	976	1197
2002	1022	1259
2003	1056	1288

Source: *New Earnings Survey*, various years

Activity 5   *Changes in the earnings ratio over time*

(a)   Calculate the earnings ratio at the mean for each year in Table 4.

(b)   How has the earnings ratio changed since 1989?

(c)   On the evidence of your calculations, would you say that gender inequalities in earnings have widened, narrowed or stayed the same between 1989 and 2003?

2.3   *Averages—the mean or the median?*

So far in investigating men's and women's earnings you used the mean when comparing levels of earnings. But there are other averages—the median, in particular. Would the results of the investigation have been the same using *median* earnings of men and women?

The mean earnings of any group of people may be thought of as the 'average' of the earnings of all the people in that group. The median earnings may be thought of as the earnings of the 'average person' in a particular sense — if you put the people in order of their earnings, then the 'average person' is exactly in the middle (that is, 50% along). So mean earnings and median earnings are different ways of measuring the 'middle' level of earnings of the group.

You may find it useful to review your notes on *mean* and *median* in the *Unit 2* Handbook activity. Are you clear about the difference between them?

Table 5 gives values of the median and mean gross weekly earnings including and excluding overtime, and the median and mean gross hourly earnings excluding overtime for adult men and women in full-time employment in 2003.

Table 5 Some 2003 earnings data

| | Median | | Mean | |
	Women	Men	Women	Men
Gross weekly earnings incl. overtime (in £)	339	432	396	525
Gross weekly earnings excl. overtime (in £)	330	398	389	500
Gross hourly earnings excl. overtime (in p)	892	1024	1056	1288

Source: *New Earnings Survey*, 2003, Tables A28, A37 and A39

The *earnings ratio at the median* is defined in a similar way to the earnings ratio at the mean, as follows.

> The *earnings ratio at the median* is: $\dfrac{\text{median earnings of women}}{\text{median earnings of men}}$.

Activity 6 *The earnings ratio at the median*

(a) Calculate the earnings ratio at the median using the data in Table 5 for each of the following: gross weekly earnings including overtime, gross weekly earnings excluding overtime, gross hourly earnings excluding overtime.

(b) Compare the three earnings ratios at the median that you calculated in part (a) with the corresponding earnings ratios at the mean (which were calculated in Example 1 and Activity 4). This latter set of values is 75%, 78% and 82%, respectively. What do you notice?

The comments on this activity (page 66) observe that in each case the earnings ratio at the median is greater than the earnings ratio at the mean. Remember the nearer any earnings ratio is to 100%, the closer the earnings of women are to those of men. So the relative 'gap' between the earnings of the 'average' man and the 'average' woman (that is, at the median) is less than that between the 'average' earnings of all men and the 'average' earnings of all women (that is, at the mean).

Looking again at Table 5, you can see that, for both men and women, the median earnings figure is less than the mean earnings figure. In fact, for earnings data, it is generally true that the median is smaller than the mean. Why should this be so?

Here is an example of a typical *distribution* of earnings; that is, how earnings vary between employees. Imagine a small manufacturing company, for instance. The earnings of the majority of the employees will probably not be very different from one another: maybe some will earn as much as twice the amount that others do, but not much more. However, there will almost certainly be one or two senior managers who earn very much more. This hypothetical distribution of earnings is illustrated in Figure 1.

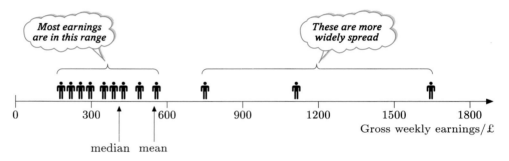

Figure 1   The distribution of earnings in a small (imaginary) company

So the earnings of the majority of employees will be fairly closely grouped, but there will be a few who earn much more. This is the case for earnings in general. This distribution of earnings is the primary reason for the phenomenon that median earnings are generally lower than mean earnings. The following activity should help you to see how this occurs.

For this activity, you need to be able to enter and edit data in the lists or statistical registers of your calculator, and to use your calculator to find the mean and the median of a list of data.

If you are not sure how to do this, then look again at Section 2.1 of Chapter 2 of the Calculator Book.

Activity 7   *A calculator investigation*

(a) Troublefree Computers has five employees, including the manager. Their weekly earnings in pounds are 300, 350, 400, 450, 500. Find the mean earnings and the median earnings of the employees.

(b) Suppose that one employee, the manager, is given a rise from £500 to £600. Calculate the mean earnings and the median earnings now. Which is the larger figure?

(c) Now vary the amount received by the manager, keeping the earnings of the other employees unchanged. Find the mean earnings and the median earnings for each of a number of different values of the manager's new salary, each larger than £500. What do you notice?

In Activity 7, you saw that as the manager's pay is increased, the mean earnings increases, but the median is unaffected. This is illustrated in Figure 2.

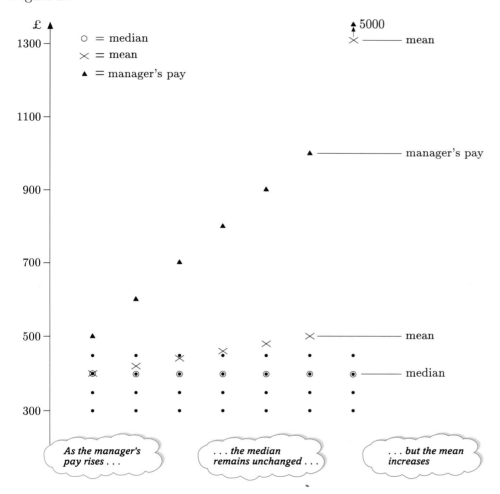

Figure 2   The mean and the median as the manager's earnings increase

The median is unaffected because the middle value remains the same however large the manager's pay becomes. The mean, however, is the total earnings of the five employees divided by five; so, as the manager's pay increases, the total earnings of the five employees increases and hence the mean rises. Increasing the manager's pay has the effect of dragging up the mean.

Since, in general, higher earnings are much more widely spread than lower earnings, this phenomenon of the mean being greater than the median is to be expected when examining earnings data. Further, the more widely spread the higher earnings are, compared with the lower earnings, the greater the difference between the mean and the median will be. For the earnings data in Table 5, the mean is greater than the median for both men and women. This reflects the fact that higher earnings are more widely spread than lower earnings for both men and women. Also, in Activity 6, you found that the earnings ratio at the median is greater than the earnings ratio at the mean in each case. This is due to higher earnings being more widely spread, compared with lower earnings, for men than for women.

Decisions about which measure to use have implications for how the data are interpreted. Since mean earnings are generally higher than median earnings, a trade union official might use median earnings to support an argument that average pay is low. On the other hand, the employer might use the mean to argue the opposite. The relatively large earnings of a minority can have a marked effect on mean earnings, whereas the median is unaffected by a few extremely large values. So there is a case for regarding the median as more representative of earnings in general than the mean. For the rest of the investigation into the earnings of men and women here, we use the median rather than the mean.

The pay parade

The following image appeared in a journal article in 1994. Imagine a parade of all the workers in the UK in which everyone's height is proportional to their weekly earnings: so a person earning an average (that is, mean) wage is of mean height. The shortest, that is the lowest-paid, pass by first, the tallest, that is the highest-paid, last. Suppose the parade takes an hour with everyone moving at the same speed. For the first twenty-five minutes all you see are very short people—ten million people less than four feet tall. Only in the last twenty minutes do you see people of average height, followed by a few giants: government ministers eight metres tall and heads of companies as tall as a skyscraper.

This idea of a 'pay parade' dates back at least to a 1971 publication by the Dutch economist Jan Pen. The figures in the text (but not the diagram itself) have been updated to relate to the UK position in 2003.

Source: *The New Review*, No. 24

Activity 8   *The pay parade*

Suppose that the parade begins at 10 am and ends at 11 am.

(a) At what time would a person of median earnings pass by?

(b) According to the description above, at what time does a person earning the mean wage pass by? What percentage of people earn less than the mean wage?

(c) Think about the cartoon image itself: what were some of your reactions to it? Were there any aspects that confused you or where you felt you were being misled?

Let's summarize the investigation so far. Several factors that affect earnings have been taken into account. Since more women than men work part-time, the investigation was restricted to full-time workers. Only workers on adult pay rates were considered. Since men work more overtime on average than women, overtime was excluded. Since the normal basic working week is slightly longer on average for men than for women, the average hourly earnings of men and women were compared instead of the average weekly earnings. Since a few well-paid individuals can strongly influence the mean but not the median, the median was used for comparisons. This is summarized in Table 6.

Table 6   Adjustments made in order to compare 'like with like'

Perceived problem	Proposed solution
More women than men work part-time.	Look only at full-time workers on adult pay rates.
Men work more overtime.	Exclude overtime.
Men work a longer basic working week.	Compare hourly earnings.
A few highly-paid individuals can seriously influence the mean.	Compare median earnings.

Even after taking all these factors into account, the earnings ratio at the median for 2003 was 87%. So it appears that, the average, adult women working full-time are paid about 87% of the amount paid to the average man for an hour's work. Does this mean that women and men are not receiving equal pay for equal work? Or are there other important factors in determining pay that have not yet been taken into account?

Perhaps the most important factor that has not yet been considered is actual occupation; this is investigated in the next section.

Outcomes

After studying this section, you should be able to:

◇   interpret data accurately from tables (Activities 2 and 3);

◇   calculate the earnings ratio at the mean and at the median given relevant data, and interpret these values (Activities 4, 5 and 6);

◇   draw general conclusions from earnings ratio calculations given appropriate data (Activities 6 and 7);

◇   explain why, for earnings data, the mean is generally greater than the median (Activities 7 and 8);

◇   start to identify when and how you can use mathematical and statistical ideas to specify and investigate a problem.

3 How does occupation affect earnings?

Aims   The main aim of this section is to investigate the link between occupation and men's and women's earnings, and to introduce some techniques for describing the *distribution* of a batch of data.   ◇

3.1 Gender and occupation

The apparent difference between men's and women's earnings could be due to a number of reasons, including the following:

◇   women being paid less in the same occupation;

◇   women being employed predominantly in occupations which have relatively low pay;

◇   women not receiving 'equal pay for work of equal value';

◇   fewer women than men in senior, more highly-paid jobs.

The first two of these will be the main focus in the next three sections. The other two are largely beyond the scope of the investigation in this unit.

So how could you set about investigating whether the difference between men's and women's earnings, observed in Section 2, is due to women being paid less than men in the same occupation, or to women being employed predominantly in lower-paid occupations?

Activity 9   *Desirable data*

If you were planning to investigate the relationship between occupation and men's and women's earnings, what sort of data would you need, and why? Think about this for a minute or two, then write down your ideas.

Many activities in this course ask you to respond by writing your ideas.

▶   What is the role of writing in learning mathematics, particularly in this course?

Writing things down can help you to form and clarify your ideas, and thus help you to learn. You need to write to answer assignment questions, and to make notes as you work through the units. The Handbook activities require clear explanations of terms in your own words.

Writing is a skill that can be improved and learned. Consider a few points to help improve your skill.

When involved in a writing task, you may find it useful to think about these questions:

◇ Why are you writing? (Apart from the obvious 'because the question asked me to'!)

◇ Who do you feel you are writing for?

◇ What effect is your writing having on your thinking about the topic?

◇ How does writing help you understand or consolidate your ideas?

As you read and write about gender and occupation in this unit, think about the characteristics of good mathematical writing.

What data are available?

The *New Earnings Survey* publishes the numbers of men and women that were included in the survey from each occupational group. The sample of employees included in the survey is assumed to be representative of all employees in the whole of Great Britain, so these numbers can be used to provide a good indication of the proportions of men and women employed in the various occupational groups listed in the survey. They will show whether certain jobs are done predominantly by men and others predominantly by women. Data are also available on both the weekly earnings and hourly earnings of men and women, either including or excluding overtime, for each occupational group.

▶ Should you use data on weekly earnings or hourly earnings?

There are advantages to each. On the one hand, as has already been observed, men tend to work more hours per week than women, on average, and using hourly earnings eliminates the effect of this factor from our investigation.

On the other hand, in many occupations, there is no paid overtime: pay is based on a nominal number of contracted hours and is fixed regardless of the number of hours actually worked, so that the published figure for hourly earnings may bear little relationship with reality. This is the case in many professions.

In secondary school teaching, for example, the basic working week according to the *New Earnings Survey* is a little over thirty hours. But according to a 2000 survey carried out by the government's Office of Manpower Economics, teachers in secondary schools actually worked about fifty-one hours each week during term time. In such occupations, *New Earnings Survey* figures for hourly earnings are fairly meaningless and should certainly not be used in comparisons with earnings in other occupations.

Taking all these factors into account, the course team settled for using data on weekly earnings excluding overtime in most cases, where the number of hours worked by men and women differ little. However, data for *hourly* earnings excluding overtime are used where the numbers of hours worked by men and women differ greatly within a given occupation.

As mentioned above, the *New Earnings Survey* includes information on the numbers of men and women surveyed in the various occupational groups included in the survey. So it is a relatively simple matter to find out whether there is evidence that women tend to work in lower-paid occupations.

First, look at the numbers of men and women in the survey who were working in the broadest groupings of occupations for which information is published. (These groups are known as Major Occupational Groups.) The figures are given in Table 7. 'Elementary Occupations' are 'occupations which require the knowledge and experience necessary to perform mostly routine tasks, often involving the use of simple hand-held tools, and in some cases, requiring a degree of physical effort' (Standard Occupational Classification, 2000, Volume 1, page 261). A typical occupation in this group is 'Labourers in building and woodworking trades'.

Table 7 Numbers of men and women surveyed in major occupational groups

Major Occupational Group	Women	Men
Managers and Senior Officials	4 962	12 290
Professional Occupations	5 642	8 585
Associate Professional and Technical Occupations	6 588	8 987
Administrative and Secretarial Occupations	12 818	5 241
Skilled Trades Occupations	628	9 673
Personal Service Occupations	3 373	1 343
Sales and Customer Service Occupations	2 981	2 434
Process, Plant and Machine Operatives	1 319	8 932
Elementary Occupations	2 504	8 603
Total	40 815	66 088

Source: *New Earnings Survey*, 2003, Tables D14 and D15

Activity 10 *Some basic facts contained in the table*

What percentage of the total number of women surveyed was in each of the occupational groups listed? Find the corresponding percentages for men. What do these percentages tell you about the occupations of men and women? Also, to what extent does placing these figures in a table make it easier for you to process them?

The data indicate that larger proportions of women than of men work in occupations classified as 'administrative and secretarial', 'personal service' and 'sales and customer service', while greater proportions of men than of women work as 'managers and senior officials', 'process, plant and machinery operatives', or in 'skilled trades' or 'elementary occupations'. The pie charts in Figure 3 represent the data in Table 7. The areas of the circles are proportional to the numbers of men and women surveyed.

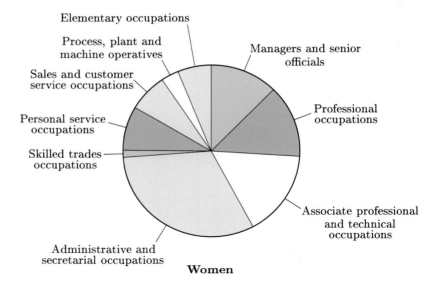

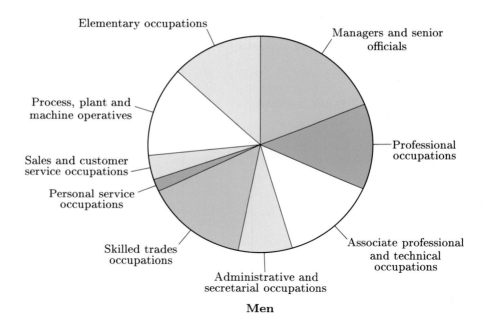

Figure 3   Pie charts showing the proportions of men and women surveyed in the major occupational groups

Each of the major occupational groups is subdivided into narrower occupational groups, and the *New Earnings Survey* published data on many of these occupations. Table 8 shows the numbers of women and men included in the survey in a variety of occupations. Even this small sample of occupations illustrates quite well that there are some jobs that are done mainly by men (for example, electrical trades) and others mainly by women (for example, nursing).

Table 8   Numbers of women and men in the survey in various occupations

	Women	Men
Chefs and cooks	207	361
Filing and records assistants and clerks	847	420
Chartered and certified accountants	186	344
Secondary education teachers	1161	938
Nurses	1895	341
Cleaners and domestics	471	496
Electrical trades	72	1722

Source: *New Earnings Survey*, 2003, Tables D14 and D15

So it is possible that the difference between men's and women's earnings might, in part, be due to occupation. The next step is to compare earnings of men and women in the same occupation to see if there is evidence that women do earn less than men within the same occupation.

Table 9 shows the median gross weekly earnings, excluding overtime, of adult men and women working full-time in six different occupational groups. The earnings ratio at the median is also shown for each group.

Table 9   Median gross weekly earnings excluding overtime (in pounds)

Occupation	Women	Men	Earnings ratio at the median (%)
Chefs and cooks	231	288	80
Filing and records assistants and clerks	275	333	83
Chartered and certified accountants	546	662	82
Secondary education teachers	571	633	90
Nurses	435	463	94
Cleaners and domestics	201	230	87

Source: *New Earnings Survey*, 2003, Tables D13 and D13a

A check on the basic hours worked was carried out: the data are given in Table 10. (The figures given are averages; that is, means.)

Table 10   Normal basic hours (averages) for women and men

Occupation	Normal basic hours	
	Women	Men
Chefs and cooks	38.0	41.4
Filing and records assistants and clerks	37.0	37.9
Chartered and certified accountants	36.6	37.1
Secondary education teachers	32.2	32.3
Nurses	37.4	37.8
Cleaners and domestics	37.5	40.1

Source: *New Earnings Survey*, 2003, Tables D1 and D2

Look at the normal basic hours worked by men and women in the occupations given in Table 10.

▶   Do you think any of the differences are large enough to account for much of the difference in earnings between men and women in these occupations?

The data in Table 10 show that for secondary education teachers, there was very little difference in the official number of hours worked by men and women. In some other groups (for example, accountants), males had slightly longer average working weeks (between half an hour and an hour longer) than females in the same occupations.

However, the difference in working hours was quite large in two of the occupational groups. The basic working week of chefs and cooks was three and a half hours longer for men than for women. For cleaners and domestics, the normal basic hours worked were two and a half hours longer for men than for women. These differences might well account for a large part of the difference in median weekly earnings of men and women in these two occupational groups. It is worth looking at the median hourly earnings for these two groups.

Table 11 shows the median gross hourly earnings (excluding overtime) for men and women in these two occupational groups.

Table 11   Median gross hourly earnings excluding overtime (in pence)

	Women	Men
Chefs and cooks	600	670
Cleaners and domestics	544	569

Source: *New Earnings Survey*, 2003, Tables D13 and D13a

Activity 11   *The median hourly earnings*

Calculate the earnings ratio at the median, based on gross hourly earnings, for men and women in both occupational groups in Table 11. State the effect that using hourly earnings data instead of weekly earnings data has had on the earnings ratio at the median.

Using hourly earnings data instead of weekly earnings data increased the earnings ratio at the median for both occupational groups.

In summary, the earnings ratios at the median in Table 9 indicate that women in these occupations are paid less than men per week. A brief investigation into hours worked has suggested that the figures might be misleading for two of the occupational groups: chefs and cooks, cleaners and domestics. In Activity 11, you calculated the earnings ratios at the median for these two groups based on hourly earnings data: these were still less than 100%. In all six occupations considered, women are paid less than men on average. For some groups, like cleaners and domestics, the difference is quite small. For others, like accountants, the difference may be as much as 18%.

So, despite the Equal Pay Act, for the occupations looked at in our investigation, it appears that women are still paid less than men on average.

Activity 12   *Finding and communicating an explanation*

Assuming that employers are not breaking the law and paying different wages to their male and female employees, can you suggest a possible explanation for this? Write a sentence or two giving your views and explanation.

3.2   *The distribution of earnings*

For each of the six occupations considered in Subsection 3.1, the data indicate that women are paid less than men on average. But that does not imply, for example, that all female accountants are paid less than all male accountants, and all female nurses are paid less than all male nurses. Accountants do not all earn exactly the same as each other, and nor do nurses. Using average earnings simplifies comparisons, but to gain a greater understanding of how women's earnings compare with men's earnings, you need to look at the *distribution* of earnings.

Some data were available in the *New Earnings Survey* on the distribution of earnings within occupations. These data are all in the form of *summary statistics*; that is, numbers which summarize the data, such as the mean and the median. Several other summary statistics are included. However, first you need to know what these statistics are, and how they are calculated.

There are several different ways of measuring the distribution and spread of earnings within a group. To illustrate them requires data on the earnings of individuals, not just summary statistics, but unfortunately such individual data are not readily available. However, the *New Earnings Survey* contained a good deal of information about the general distributions of earnings within occupations. This has been used to construct some sets of data to illustrate points in this subsection.

The data in Table 12 on the earnings of twenty-two chartered and certified accountants (eleven male and eleven female) were made up from information on the distribution of earnings contained in the 2003 *New Earnings Survey*. These data will be used to introduce you to two of the different measures of *spread*.

Other measures of spread will be introduced later in the course.

Table 12   The gross weekly earnings of twenty-two chartered and certified accountants (in pounds)

Women	319	418	431	503	531	604	631	654	865	872	1019
Men	376	392	536	542	618	649	702	876	881	1016	1117

The range

The *range* of a batch of data is the numerical difference between the smallest or minimum value, called *min*, and the largest or maximum value, called *max*.

> range = $max - min$.

Activity 13   Calculating the range

Find the range of the batch of women's earnings in Table 12. Find the range of the batch of the eleven men's earnings. Is the range greater for the women or for the men?

Quartiles and the interquartile range

Given a batch of data in numerical order, such as either of the two batches in Table 12, the range can easily be calculated, but it reveals nothing about how the values in the main body of the data are distributed. For example, the values could be bunched around the median, or they could be spread fairly evenly between the minimum and maximum values, *min* and *max*.

The range is also very sensitive to changes in the two extreme values, *min* and *max*. Suppose, for example, that the highest-paid of the female accountants was an exceptionally talented and successful woman earning, say, £2500 a week, instead of £1019. Then the range of the batch would have been £(2500 − 319) = £2181, giving the impression that the earnings of the women accountants were very widely spread. In fact, apart from this one value, all the values in the batch were within £553 of each other. A measure of spread which is not influenced by extreme values in this way would be useful.

What is needed is a measure which conveys information about the spread of values *in the main body of the data*.

Activity 14   Assessing some measures of spread

Here are three possible measures of spread. What are some strengths and weaknesses of each measure? Can you suggest an alternative measure?

(a) Exclude the two extremes and find the range of the remaining values.

(b) Exclude four values from each end of the batch and find the range of the remaining values.

(c) Include only the first four values each side of the median and find the range of these values.

What is required is a measure that takes into account the size of the batch of data. One such measure is based on the difference between two particular values in the batch, known as the *quartiles*. As the name implies, the two quartiles lie approximately one quarter of the way into the batch from either end. Roughly speaking, 25% of the values in a batch lie below the *lower quartile* and 25% of the values lie above the *upper quartile*: roughly speaking, 50% of the values lie between the quartiles. This is illustrated in Figure 4.

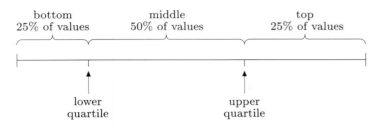

Figure 4   The upper and lower quartiles of a batch of values

Notice the use of the word 'approximately' and the phrase 'roughly speaking'; to see why they are there, look at this example. Suppose that there are 10 numbers in a batch. Now 25% of 10 is 2.5, which is not a whole number. So not exactly 25% of the values may lie below the lower quartile.

There are a number of subtly different criteria for deciding on the exact values of the quartiles: the criterion described in this course is not the only one. You may well come across slightly different ones in another course or in a book or when using a statistics package on a computer—therefore, you should be aware that there are other possibilities. So small differences in the actual values of the quartiles may occur. However, whichever method is used, the results obtained are similar.

In the calculator work that follows, you will be asked to investigate the rules that your calculator uses to find the quartiles of a batch of data.

Now work through Section 3.1 of Chapter 3 of the Calculator Book.

Finding the quartiles of a batch is very similar to finding the median. First, sort the batch into order, smallest first. Then the median is the middle value, or the midpoint of the two middle values if there is an even number of values in the batch.

The *lower quartile*, which is denoted $Q1$, is the median of the lower half of the batch of data—that is, of the values to the left of the median in your list. The *upper quartile*, which is denoted $Q3$, is the median of the upper half of the batch of data—that is, of the values to the right of the median.

With this new labelling, it makes sense, retrospectively, to think of the median as $Q2$, though this is not standard notation.

Here is a notational device that may help you picture these ideas. Imagine the values in the batch spread out in a line—you might want to think of them as little balls held together by sticks. Imagine taking away the middle point—the median—so that the line of data splits into two halves at the median. Now imagine picking out the middle point of the lower half—the values *below* the median—this point will be *Q1*. Finally, imagine picking out the middle point of the upper half—this is *Q3*. For example, for a batch of size seven, the configuration would look like Figure 5.

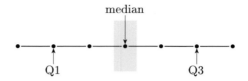

Figure 5   Line diagram for a seven-item batch

In this case, there is an odd number of values in the batch, so the median is the middle value. There are three values in the lower part of the batch, to the left of the median. The lower quartile is the middle one of these three values. The upper quartile is the middle one of the three values to the right of the median.

For a batch of size eight, the diagram looks like Figure 6.

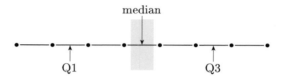

Figure 6   Line diagram for an eight-item batch

In this case, there is an even number of values, so the median is halfway between the two middle values. None of the eight values in the batch lies exactly at the median. Hence there are four values in the lower half (to the left of the median). The lower quartile is the median of these four values, and is therefore halfway between the second and third values. The upper quartile is found in a similar way by finding the median of the four values in the upper half of the batch.

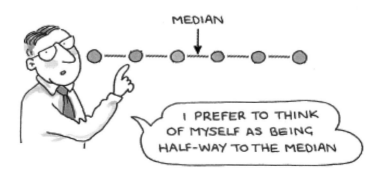

Activity 15   *Drawing diagrams*

(a) Draw a line diagram to represent a batch of nine values and mark on it the positions of the median and quartiles.

(b) Draw a line diagram to represent a batch of ten values.

Activity 16   *Using diagrams*

(a) Draw a line diagram to represent a batch of eleven values.

(b) Use your diagram to find the median and quartiles of the batch of men's earnings in Table 12 on page 28.

(c) Use your diagram to find the median and quartiles of the batch of women's earnings in Table 12 (page 28).

There is a measure of spread based on the quartiles.

> The *interquartile range* is the numerical difference between the upper and lower quartiles: $Q3 - Q1$.

Example 2   *Interquartile range*

Find the interquartile range for the batch of men's earnings in Table 12.

$Q3 - Q1 = £881 - £536 = £345.$

Activity 17   *Finding the interquartile range*

Find the interquartile range for the batch of women's earnings in Table 12. Is this greater or smaller than the interquartile range for the men's batch?

Activity 18 *Comparing the spreads*

(a) For the data given below in Table 13, find the range and interquartile range of the men's earnings and of the women's earnings. (Use your calculator to sort the data into ascending order.)

(b) Briefly compare the spread of the earnings of male and female chefs and cooks.

Table 13 The gross hourly earnings excluding overtime of seventeen chefs and cooks (in pence)

Women	536	495	667	503	616	689	579	1010	
Men	665	634	818	531	686	611	774	981	963

Two *measures of spread* (the range and the interquartile range) have been introduced in this subsection; these have been used to compare the spread of batches of earnings of men and women in two occupations (chartered and certified accountants; chefs and cooks). In both cases, one measure of spread was greater for the men than for the women and it was the other way round for the other measure. Although, for these batches, some of the women earned more than some of the men, the earnings of the men were generally higher than the earnings of the women.

Outcomes

After studying this section, you should be able to:

◇ extract relevant information from tables of data (Activities 10, 11, 13, 17 and 18);

◇ describe some of the advantages and disadvantages of different measures of spread (Activity 14);

◇ explain the meaning of the terms 'range', 'lower quartile', 'upper quartile' and 'interquartile range' and find the values of each for a given batch of data (Activities 13, 16, 17 and 18);

◇ find the median and the quartiles of a batch (Activities 15 and 16);

◇ produce short written descriptions and explanations exploring certain economic and social issues (Activity 12).

4   Boxplots

Aims   The main aim of this section is to introduce the boxplot as a way
of representing the distribution of a batch of data and to show how
boxplots can be used to compare batches of data.   ◇

In the previous section, you saw that, in each of the six occupations looked
at, the median earnings of men was higher than the median earnings of
women. This prompted the question: 'Is there a difference between men's
and women's earnings across the whole range from the lowest-paid to the
highest-paid?'

The median and the quartiles were used to provide information about the
level of earnings in a batch, and the range and interquartile range were
used to measure the spread of earnings. In this section, a diagram which
illustrates this information is introduced: the boxplot. Its major strength
is that it allows you to make a direct visual comparison between two
batches of data. As you probably noticed in *Unit 2*, it is often easier to
interpret information which is presented in a diagram rather than as a list
of numbers.

4.1   What is a boxplot?

There are five key values which can summarize the distribution (or shape)
of a batch of data: the two quartiles ($Q1$ and $Q3$), the median (which is
actually $Q2$), and the two extreme values (min and max).

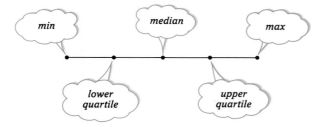

Figure 7   The five key values

These key values can be shown on a line diagram, but another useful
diagram is called a *boxplot*.

For the earnings values of male accountants given in Table 12 (page 28), the boxplot looks like Figure 8.

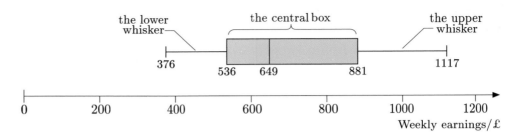

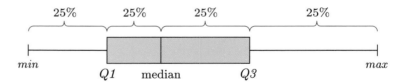

Figure 8   Boxplot of earnings of eleven male accountants

The central feature of a boxplot is a *box* which extends from the *lower quartile* to the *upper quartile*. This part of the diagram contains approximately 50% of the values in the batch. The length of the box is equal to the *interquartile range*.

Boxplots are sometimes called *box and whisker diagrams*.

Outside the *box* are two *whiskers* extending from each quartile to the corresponding extreme value. Each whisker covers 25% of the remaining batch. The distance between the extreme ends of the two whiskers is equal to the *range*.

The median is shown on a boxplot by putting a vertical line through the box. Since the median may be nearer one quartile than the other (as in Figure 8), the vertical line representing it will often not be centrally placed in the box.

Thus, a boxplot shows clearly the division of a batch of data into four parts: the two whiskers and the two sections of the box. Each contains (approximately) 25% of the values in the batch, as shown in Figure 9.

Figure 9   A typical boxplot

The boxplot in Figure 9 is a fairly typical one in shape: most batches of data are more densely packed in the box (that is, the middle of the batch) and less dense near the extreme values.

Example 3   Boxplot for female accountants' earnings

Figure 10 shows the boxplot for the gross weekly earnings of the eleven female accountants given in Table 12 (page 28).

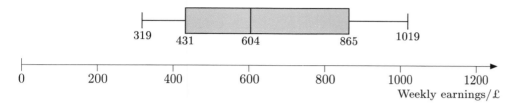

Figure 10   Boxplot of earnings of eleven female accountants

As you can see, the whiskers are shorter than the corresponding sections of the box, indicating that the values are less widely spread out in the whiskers than in the box.

Notice that, in both Figure 8 and Figure 10, the five values—the median, the two quartiles and the two extremes—are written on the boxplot. This is customary, so remember to include them whenever you draw a boxplot.

4.2   Comparing batches using boxplots

A boxplot can be a very effective way of presenting information about the distribution of a batch of data in a diagram. However, it really comes into its own when used for comparing two or more batches of data. If boxplots for two batches are drawn on the same diagram, with the same scale, then a direct visual comparison of the two batches can be made.

Figure 11 shows the boxplots for the male and female accountants of Table 12, on the same diagram.

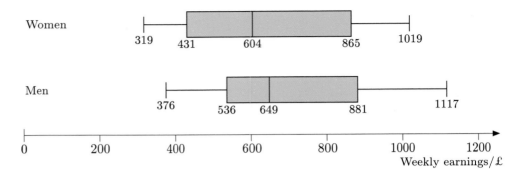

Figure 11   Gross weekly earnings of accountants

Example 4   *Comparing earnings using boxplots*

Use the boxplots in Figure 11 to compare the earnings of the male and female accountants.

The main message from Figure 11 is that *all* the key points on the boxplot—the median, the quartiles and the extremes—are higher for the men than the corresponding values for the women. This shows that the level of earnings for the men is generally higher than that for the women.

Since in Figure 11 the whiskers are longer in the men's boxplot than in the women's boxplot, the earnings of the top 25% and bottom 25% of earners are more widely spread for the men than for the women.

There is one other point worth noting about the batches of earnings data shown in Figure 11. The batches are not symmetric: for each boxplot, the right-hand whisker is longer than the left-hand whisker, and the right-hand section of the box is longer than the left-hand section: the boxplots are *asymmetric*. This reflects the fact that the higher earnings values are more widely spread than the lower values. This is frequently the case with earnings data: remember the image of the pay parade (Activity 8 in Section 2).

This example shows that it is relatively easy to compare batches of data by looking at boxplots (drawn on the same diagram with the same scale). Trying to make the same comparisons from tables of individual data values is much harder.

A batch of data which exhibits the type of asymmetry described in Example 4—one whisker of the boxplot and the corresponding section of the box being longer than the other—is said to be *skewed*. Earnings data are generally *right-skewed*, since a boxplot of earnings data usually exhibits a long *right* whisker and *right*-hand section of the box. Whenever the boxplot for a batch of data has a long right whisker, the mean is usually larger than the median. In particular, the mean of a batch of earnings data will generally be higher than the median. (Recall the discussion about the mean and the median of earnings data in Section 2.)

WHEN ONE WHISKER IS LONGER
THAN THE OTHER,
THE DISTRIBUTION IS SKEWED

Of course, not all batches of data, not even all batches of earnings data, are right-skewed. A batch may be symmetric—in this case, the two whiskers will be roughly the same length and the two sections of the box will be roughly the same length (as in Figure 12(a)). Or a batch may be *left-skewed*—this is the case when the left whisker is longer than the right whisker and the *left-hand* section of the box is longer than the right-hand section (as in Figure 12(b)).

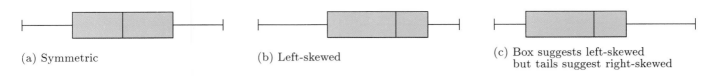

(a) Symmetric   (b) Left-skewed   (c) Box suggests left-skewed
                                                                         but tails suggest right-skewed

Figure 12   Three boxplots

Of course, the boxplot for a batch of data may not fit any of these descriptions; for example, it may have a longer right whisker than left, but a longer left-hand section of the box (as in Figure 12(c)). In such a situation, when describing the shape of the data, make it clear whether you are talking about the box or the whiskers of the boxplot.

Now work through Section 3.2 of Chapter 3 of the Calculator Book.

Activity 19   Comparing earnings for chefs and cooks

A sketch of the boxplots for the chefs' and cooks' earnings data from Table 13 (page 32) is shown in Figure 13.

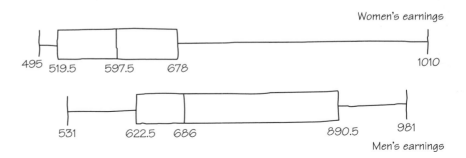

Figure 13   Sketch of boxplots of earnings data for chefs and cooks

(a) What do the boxplots tell you about the relative earnings of male and female chefs and cooks?

(b) What do the boxplots tell you about the shapes of the distributions of the earnings of male and female chefs and cooks?

(c) Make some observations about how boxplots helped you to interpret the data presented.

Detailed information and advice on the drawing of boxplots is in the Appendix 'Drawing accurate boxplots' on page 61. If you are happy about how to draw an accurate boxplot by hand (including a scale, as in Figure 11), just go straight on to Activities 20–22 now. Otherwise, you may wish to study the material in the Appendix in more detail first.

Activity 20   Drawing accurate boxplots

Sketches of the boxplots of the chefs' and cooks' earnings data of Table 13 (page 32) are shown in Figure 13. Choose a suitable scale and then draw the axis and accurate boxplots.

Table 14 contains earnings data for a small number of male and female nurses, and for male and female secondary education teachers. In the next two activities, you are asked to use these batches of data to compare the weekly earnings of men and women in these occupations. This is to provide you with more practice at using your calculator to obtain boxplots, drawing rough sketches of and accurate diagrams of boxplots, and interpreting the information contained in them.

Table 14 Weekly earnings (in pounds)

Nurses										
Women	325	450	753	434	472	484	546	449	579	440
Men	303	264	781	468	572	553	507	415	446	451	513

Secondary education teachers										
Women	454	594	258	372	612	588	604	997	519	
Men	1020	614	946	572	883	663	379	634	662	677

Activity 21 *Drawing and interpreting boxplots*

(a) Use your calculator to obtain boxplots of the nurses' earnings data in Table 14, and hence draw a rough sketch of the boxplots.

(b) Draw your own accurate diagram of the boxplots.

(c) What do your boxplots tell you about the relative earnings of male and female nurses?

Activity 22 *More drawing and interpreting boxplots*

(a) With the aid of your calculator, sketch boxplots for the earnings data in Table 14 for male and female secondary education teachers.

(b) Draw your own accurate diagram of the boxplots.

(c) What do your boxplots tell you about the relative earnings of male and female secondary education teachers?

When the boxplots for two batches of earnings data are drawn on the same diagram to the same scale, a visual comparison of the level and distribution of the earnings in the two batches can be made. You have probably found it easier to compare two boxplots than to compare two lists of numbers. In this section, boxplots have been used to compare the distribution of the earnings of men and women in four occupational groups. In three cases, they showed that the level of the men's earnings was generally higher, although there was little difference for the nurses (see Activity 21).

Outcomes

After studying this section (and possibly the Appendix), you should be able to:

◇ compare batches of data by interpreting boxplots (Activities 19, 21 and 22);

◇ use your calculator to display a boxplot;

◇ identify the median, the lower quartile, the upper quartile, the smallest and largest values of a batch of data, given a boxplot (Activity 19);

◇ draw a rough sketch of a boxplot (Activities 21 and 22);

◇ draw a boxplot accurately by hand (Activities 20, 21 and 22).

5 Interpreting and representing earnings data

Aims   The main aim of this section is to show how boxplots are used to represent summary statistics from the *New Earnings Survey*, and to show how boxplots and earnings ratios may be used to compare the earnings of men and women in a number of occupations.   ◇

This section completes the investigation into the earnings of men and women, and in particular into the relationship between gender and occupation. The batches of earnings data used in Section 4 were small. However, the *New Earnings Survey* used much larger samples. Some of the summary statistics which are available will be described in this section.

Data on the distribution of earnings in a large number of occupational groups are published in several forms in Volume D of the *New Earnings Survey*. One set of tables gives information such as the median and the quartiles for men and women in each occupation.

Table 15   Distribution of gross weekly earnings of chartered and certified accountants in 2003 (in pounds)

	Women	Men
Upper quartile	718	817
Median	546	662
Lower quartile	461	548

Source: *New Earnings Survey*, 2003, Tables D13 and D13a

Activity 23   *Interpreting the table*

(a) What percentage of the men referred to in Table 15 earned between £548 and £817?

(b) What was the largest amount earned by any of the lowest-paid 25% of the women?

(c) Approximately what percentage of the men earned more than £817 during the week of the survey? Approximately what percentage of the women earned less than £718 per week?

Deciles

The *New Earnings Survey* also gave information on deciles. The lowest and highest *deciles* of a batch of data provide information about the tails of its distribution. Just as the quartiles cut off a quarter, or 25%, of the values at either end, the *highest decile* cuts off the top tenth or 10% of values, while the *lowest decile* cuts off the bottom 10%. This is illustrated in Figure 14.

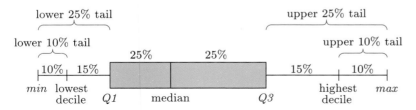

Figure 14   The highest and lowest deciles on a boxplot

Table 16 includes these summary statistics for male and female accountants in addition to the information given in Table 15. (Note that the *max* and *min* earnings are not published in *New Earnings Survey* reports.)

Table 16   Distribution of gross weekly earnings of chartered and certified accountants in 2003 (in pounds)

	Women	Men
Highest decile	874	1061
Upper quartile	718	817
Median	546	662
Lower quartile	461	548
Lowest decile	374	448

Source: *New Earnings Survey*, 2003, Tables D13 and D13a

Example 5   *Interpreting summary statistics*

Interpret the figures in Table 16 for male accountants.

The median earnings of male chartered and certified accountants was £662. This means that, in the week of the survey, 50% of this group of male accountants earned more than £662 while 50% earned less than £662. The upper and lower quartiles in this column are £817 and £548 respectively, so 25% of this group earned less than £548 and 25% earned more than £817. The extra information obtained by looking at the highest decile (£1061), and the lowest decile (£448), is that 10% of this group earned more than £1061 and 10% earned less than £448. Note that you are not told the values of the highest and lowest amounts earned by any individual in this group.

Of course, if one of the men earned exactly £662, the median earnings, then the percentage earning more than £662 would not be exactly 50%, but only approximately 50%. Similar comments apply if someone's earnings are exactly equal to the lowest decile, or the lower quartile, etc.

Activity 24 Interpreting Table 16

(a) What percentage of the female accountants included in the survey earned £874 or more?

(b) What percentage of the female accountants earned £374 or less?

(c) What percentage of the female accountants earned between £546 and £874?

(d) What percentage earned between £374 and £461?

Decile boxplots

As you saw in Section 4, it is usually very much easier to grasp the distribution of a batch of data or to compare two batches of data by looking at boxplots than by studying lists of numbers. It would therefore be informative to draw a boxplot for the type of data in Table 16. However, the boxplot cannot be exactly like those drawn in Section 4, since the extremes of a batch, *min* and *max*, which were marked on the boxplot, are not given in Table 16.

Recall that if the extreme values in a batch are unusually large or small, including them may give a distorted picture of the main body of the data: this was the reason for preferring the interquartile range to the range as a measure of spread. When dealing with large batches of data from surveys like the *New Earnings Survey*, it is common not to be given the extreme values, but to be given the highest and lowest deciles instead. One reason for this is that the deciles are not as affected by unusual values in the way that the extremes are. So a boxplot is often drawn which extends only from the lowest decile to the highest decile. This is called a *decile boxplot*.

Example 6 Decile boxplots for accountants' earnings

Figure 15 shows decile boxplots for the weekly earnings of the male and female accountants included in Table 16.

Notice that arrowheads are used at the ends of the whiskers instead of vertical bars to indicate that these points represent the highest and lowest deciles and *not* the extremes. The arrowheads point to the missing 10% of the values on either side. The *tip* of each arrow is at the appropriate decile.

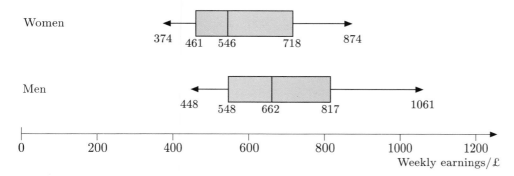

Figure 15   Decile boxplots of accountants' earnings (data from Table 15)

Activity 25   *Interpreting decile boxplots*

What do the boxplots in Figure 15 tell you about the relative earnings of male and female accountants?

It is immediately clear from Figure 15 that the spread of the earnings is greater for the men than for the women, both within the box and in the whiskers. Earnings are higher for men than for women across the whole range.

Now consider male and female chefs and cooks. The basic working week is longer for male chefs and cooks than for female chefs and cooks. So, in order to compare like with like, use data on *hourly* earnings. Table 17 contains earnings data from the 2003 *New Earnings Survey*.

Table 17   Distribution of gross hourly earnings of chefs and cooks in 2003 (in pence)

	Women	Men
Highest decile	874	1079
Upper quartile	720	862
Median	600	670
Lower quartile	513	560
Lowest decile	469	500

Source: *New Earnings Survey*, 2003, Tables D13 and D13a

Activity 26   *Drawing and interpreting decile boxplots*

Draw decile boxplots for these data and use them to compare the levels and distributions of the earnings of male and female chefs and cooks. Write a paragraph summarizing what you have found out.

Look back to the earlier paragraphs on good writing (pages 21–22) and look at your own responses. Could another student understand your writing?

Earnings ratios

In Section 2, earning ratios at the mean and at the median were used to compare the average earnings (mean and median) of men and women. They can also be used to compare earnings over the whole range from the lowest-paid to the highest-paid. The earnings ratios at the quartiles and at the highest and lowest deciles are defined in a similar way to the earnings ratios at the mean and median. For example, the *earnings ratio at the lower quartile* is calculated as follows.

$$\frac{\text{lower quartile earnings of women}}{\text{lower quartile earnings of men}}$$

The *earnings ratio at the highest decile* is calculated as follows.

$$\frac{\text{highest decile earnings of women}}{\text{highest decile earnings of men}}$$

Recall from page 11 that earnings ratios are often expressed as percentages: to convert the figure from the above formula, merely multiply by 100% (and round to the nearest whole number).

The earnings ratios at the median, quartiles and deciles for the accountants of Table 16 are given in Table 18.

Table 18 Earnings ratios for accountants, calculated from data in Table 16 (page 41)

Highest decile	82%
Upper quartile	88%
Median	82%
Lower quartile	84%
Lowest decile	83%

These figures confirm that the earnings of female accountants are lower than the earnings of male accountants over the whole range. Remember, a low earnings ratio (considerably lower than 100%) indicates a large 'gap' (measured relatively) between women's and men's pay.

Activity 27 Earnings ratios

Calculate the earnings ratios at the median, the upper and lower quartiles and the highest and lowest deciles for the chefs and cooks in Table 17. Comment on what these figures tell you about the relative earnings of male and female chefs and cooks.

In this section, decile boxplots and various earnings ratios have been used to compare the distributions of the earnings of men and women in various occupations. Decile boxplots provided an informative way of representing visually some of the data from the *New Earnings Survey*. Earnings ratios allowed the measurement of the relative earnings of women and men at different points of the two distributions.

▶ What has been discovered about the relative earnings of men and women?

In Section 2, it was found that in 2003 women received, on average (excluding overtime), only about 85% of the amount paid to men for an hour's work.

The last three sections have been concerned with investigating whether this difference between the earnings of men and women is due to women being paid less than men in the same occupation, or to women being employed predominantly in occupations which have relatively low pay. Data on the numbers of men and women in several occupations confirmed that the proportions of male and female workers vary from occupation to occupation, so this could account for some of the difference in earnings. However, in most of the occupations for which data were obtained, the men's earnings were generally higher than the women's.

How is this to be interpreted? Does this mean that men and women do not receive equal pay for similar work? Or are there still other factors that have not been taken into account? For example, might the difference in earnings within an occupation be due to men obtaining earlier promotion than women—or to more men than women obtaining promotion overall?

These are much more difficult questions to answer and the data needed to answer them are unfortunately not readily available.

▶   So what can be concluded?

It seems that men do earn more than women on average, and at least some but not all of the difference may be accounted for by factors such as occupation and hours worked. However, a much more detailed investigation would be required to determine the reasons for the difference between the earnings of men and women *within* an occupation. As yet, no firm conclusion can be reached from this investigation about whether men and women receive equal pay for equal work.

Outcomes

After studying this section, you should be able to:
◇   interpret tables giving medians, quartiles and deciles of earnings (Activities 23 and 24);
◇   explain the terms 'highest decile' and 'lowest decile';
◇   draw decile boxplots to represent earnings data given in the form of summary statistics (Activity 24);
◇   make comparisons between distributions of earnings data using decile boxplots (Activities 25 and 26);
◇   calculate and interpret quartile and decile earnings ratios (Activity 27).

6 Changes in earnings

Aims   The main aim of this section is to discuss what the Average Earnings Index measures and how it may be used in conjunction with the Retail Prices Index to investigate the question 'Are people getting better off?'   ◇

Part of a trade union's job is to try to protect and improve the standard of living of its members. In 2004, when asked about changes in the UK statutory minimum wage, Brendan Barber (General Secretary of the Trades Union Congress (TUC)) replied, 'The TUC will be campaigning hard for further increases above inflation.' How often have you heard similar statements? You may have noticed that pay claims tend to rise and fall with inflation. People generally feel that if they are to be 'better off', then they must obtain pay rises above the increase in prices that has occurred since their last pay rise. So the investigation into the question 'Are people getting better off?' will involve comparing *changes* in prices with *changes* in earnings.

In October 2004, Philip Thornton, Economics Correspondent of the (UK) *Independent* newspaper, wrote 'Inflationary pressure is building across the economy according to reports yesterday showing a jump in factory prices and a warning of a likely surge in wage claims. The rise in prices of goods leaving the factory gate vaulted to an eight-year high last month as firms managed to pass on some of the impact of soaring oil prices. Meanwhile labour market experts warned a surge in inflation would trigger an upturn in pay deals in the crucial new year bargaining period.' (*Independent*, 12 October 2004.) So while there is always pressure from employees for pay to keep ahead of prices, there is also pressure from government and employers to restrain pay rises and keep costs down.

Factors other than inflation also play a part in determining the level of pay settlements in different occupations and companies and in different parts of the UK. One factor is how easy it is to find suitably skilled employees. The scope for productivity gains and the number of unemployed workers also have an effect. Clearly some groups of workers will obtain larger pay rises than others in any year. Some may obtain pay rises greater than the rate of inflation and some smaller. Some will become better off while others will not. The question 'Are people getting better off?' once again seems simplistic. However, changes in *average* earnings give some insight into whether, in general, (employed) people are getting better off. But, because this involves looking at averages, these conclusions will not be valid for everyone.

In *Unit 2*, you saw how the Retail Prices Index and Consumer Price Index are used to measure changes in the overall level of prices. An index of earnings, the *Average Earnings Index* (AEI), can be used to measure *changes* in earnings.

6.1 The Average Earnings Index

The Average Earnings Index (AEI) is calculated by the Office for National Statistics once a month. It measures *changes* in employed people's main source of income: their earnings. The data used to calculate the AEI come from a survey called the *Monthly Wages and Salaries Survey*. Whereas the *New Earnings Survey* covers a sample of *individuals*, this survey covers a sample of *firms*. Each month, a sample of over 8000 firms is sent a simple questionnaire.

The information obtained from each firm includes the total number of employees and the total gross amount paid to these employees (including overtime, holiday pay, and other payments). No deductions are made for income tax, or for employees' national insurance and pension contributions. Fees paid to directors are excluded. The earnings of all employees, from the lowest-paid to the highest-paid, are added together to provide the total gross amount paid by the firm. For the Average Earnings Index, *earnings* simply means *gross amounts paid to employees*.

The sample is drawn only from Great Britain, so that the AEI provides no information on Northern Ireland or on the whole UK.

There is a considerable deal of similarity between the ways of calculating the AEI, the RPI and the CPI. Recall that the goods and services whose prices are used to calculate the RPI are divided into groups, subgroups and sections. In a similar way, the firms and organizations covered by the AEI are classified according to the type of work they do: they are divided into twenty groups which are each subdivided into subgroups.

The first stage in calculating the index is to find the *average weekly earnings* for each subgroup: this represents the total gross amount paid to all employees of firms in the subgroup divided by the total number of employees of these firms. Next, the *average weekly earnings for each group* is found using a weighted mean. The *overall average weekly earnings* is a weighted mean of the average weekly earnings for the twenty groups. The weights represent the total number of employees in each group or subgroup, as a proportion of the total number of employees in the whole Great Britain economy.

The actual process of calculating average weekly earnings is rather complicated, because it has to take into account the way the firms are sampled.

Then the Average Earnings Index (AEI) is calculated by comparing the overall average weekly earnings with the corresponding figure for the base year. At the time of writing (2004), the base year is 2000 and the average AEI for 2000 is set at 100. Notice that, whereas for the RPI average prices are compared with those in January of the base year, for the AEI average earnings are compared with the average over the whole base year.

For the CPI, average prices are compared with the average over the whole base year.

Finally, the AEI figures are *seasonally adjusted*. This means that they are adjusted to allow for the effect of changes in earnings levels that occur regularly at fixed times of the year. Thus, if the AEI shows an increase between one month and the next, this means that wages have gone up (on average) that month by *more* than they would normally be expected to increase at that time of year.

The AEI provides information on changes in the overall level of earnings in Great Britain. Recent figures for the index are published each month on the National Statistics website at www.statistics.gov.uk.

Figure 16 shows the value of the Average Earnings Index for each month between January 2000 and June 2004. The plotted points have been joined by straight lines so that any patterns in the plot can be seen more clearly. (Notice that during 2000, the value of the index was less than 100 in some months and greater than 100 in others. The *average* value for 2000, the base year, was 100.)

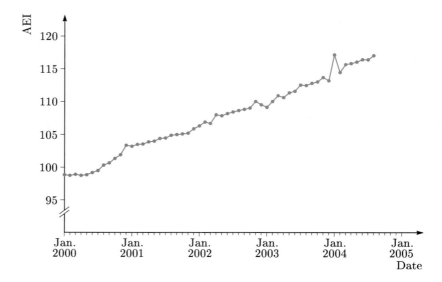

Figure 16   The Average Earnings Index, 2000–2004

Activity 28   *The peaks and dips in average earnings*

From Figure 16, it is clear that between January 2000 and June 2004 there was an upward trend in average earnings. However, average earnings did not rise steadily throughout this period: there are peaks and dips in the graph. Can you suggest any explanation for these features?

Example 7   *Using the AEI*

Estimate the percentage increase in average earnings over the year from October 2002 to October 2003.

The value of the AEI for October 2002 was 109.0 and the value for October 2003 was 113.0.

The value in October 2003 as a percentage of its October 2002 value was

$$\frac{113.0}{109.0} \times 100\% \simeq 103.7\%.$$

Thus, average earnings increased by 3.7% over the year from October 2002 to October 2003.

Activity 29   *Calculating changes in average earnings*

The value of the AEI for June 2003 was 111.6 and its value for June 2004 was 116.3. Find the percentage increase in the value of the AEI between June 2003 and June 2004.

6.2   *Prices and earnings*

The central question which motivated investigations of prices and earnings was: 'Are people getting better off?' At the beginning of *Unit 2*, you saw that this is a difficult question to answer precisely. There are many different factors to take into account, and also what may be true for one person or household may not be true for another. However, two key factors, prices and earnings, are certainly important.

In *Unit 2*, you saw how the Retail Prices Index (RPI) is used to measure changes in prices; and you have just seen how to use the Average Earnings Index (AEI) to measure changes in earnings. In this subsection, the RPI will be used together with the AEI to compare changes in prices and earnings.

To compare changes in the RPI with changes in the AEI, changes in both indices must be calculated. For each month in the period from 1990 to 2003, the percentage increase in the value of the index over its value a year earlier was calculated for both the RPI and the AEI. The results are shown in Figure 17 (overleaf).

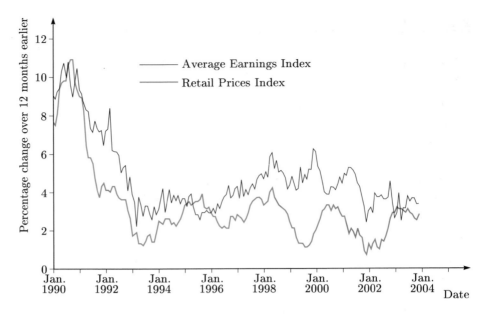

Figure 17   Changes in the AEI and the RPI: 1990–2003 (increases over previous year)

▶ What does the graph reveal about changes in prices and earnings during the period from 1990 to 2003?

First, remember that these are graphs of *increases* over a year earlier. Since the graphs remain above the horizontal axis (the level which corresponds to no change over a year earlier), both prices and earnings were rising throughout the whole period. Even though the graphs appear to be falling in the early 1990s—the *rate* of inflation and the average level of pay *increases* were falling—both prices and average pay themselves were still *rising* throughout the whole period. At no time during the period were either prices or earnings lower than they were a year earlier.

Since in Figure 17 the graph of the increase in the AEI is above the graph of the increase in the RPI for almost all of the period shown, the annual increase in average earnings was consistently higher than the annual rate of inflation: average earnings rose faster than prices. At the beginning of the period, prices and earnings were rising much more rapidly than in latter years, and for a time in 1990, prices were rising more quickly than average earnings. (This is shown by the fact that at that time, the graph of the annual increase in the RPI was above the graph of the annual increase in the AEI.) Prices were again rising faster than earnings in 1995, and briefly in 2002-03, but at these times, both prices and earnings were rising much more slowly than at the start of the 1990s.

One way to compare changes in prices and earnings is to calculate the AEI ratio as a percentage of the RPI ratio.

Example 8   *Comparing changes in prices and earnings*

(a) The value of the AEI for January 2002 was 106.3 and the value for January 2001 was 103.2.

Therefore, the value of the AEI in January 2002 as a percentage of its January 2001 value was:

$$\frac{106.3}{103.2} \times 100\% \simeq 103.0\%.$$

Thus, the increase in the value of the AEI over the year January 2001 to January 2002 was 3.0% of the January 2001 value.

(b) The values of the RPI in January 2002 and January 2001 were 173.3 and 171.1 respectively.

Thus, the value of the RPI in January 2002 as a percentage of its value a year earlier was:

$$\frac{173.3}{171.1} \times 100\% \simeq 101.3\%.$$

So average prices increased by 1.3% over the year.

(c) Now compare the changes in the two indices by calculating the ratio of ratios $\dfrac{\text{AEI ratio}}{\text{RPI ratio}}$: that is, $\dfrac{103.0}{101.3}$, and expressing this as a percentage:

$$\frac{103.0}{101.3} \times 100\% \simeq 101.7\%.$$

The AEI ratio is 101.7% of the RPI ratio. Since this ratio of ratios is more than 100%, earnings increased more than prices.

Comparisons like this give us a measure which is called the *real earnings for that month compared with one year earlier*. Real earnings for January 2002 were 101.7% of their value a year earlier.

The term 'real earnings' is used here because the ratio of actual earnings (given by the AEI for that month divided by the AEI one year earlier), divided by the corresponding ratio for prices (obtained from RPI values) gives a measure of the change in the purchasing power of earnings. To calculate this measure directly the following formula is used.

This figure is often expressed as a percentage.

> Real earnings for month x compared with one year earlier $=$
>
> $$\frac{\text{AEI for month } x}{\text{AEI for month } x, \text{ one year earlier}} \div \frac{\text{RPI for month } x}{\text{RPI for month } x, \text{ one year earlier}}$$

There is another form of this formula, which is often easier to use. Using the data in Example 8, the real earnings for January 2002 compared with one year earlier are:

$$\frac{\text{AEI for January 2002}}{\text{AEI for January 2001}} \div \frac{\text{RPI for January 2002}}{\text{RPI for January 2001}} = \frac{106.3}{103.2} \div \frac{173.3}{171.1}.$$

See Preparatory
Resource Book A.

This amounts to dividing one fraction by another. But to divide by a fraction, you turn it upside down and multiply. So the real earnings can also be calculated from

$$\frac{106.3}{103.2} \times \frac{171.1}{173.3} \simeq 1.017.$$

This can be multiplied by 100% to express it as a percentage, giving 101.7% as before. So the real earnings for January 2002 compared with one year earlier can be calculated from

$$\frac{\text{AEI for January 2002}}{\text{AEI for January 2001}} \times \frac{\text{RPI for January 2001}}{\text{RPI for January 2002}}.$$

You will not have to
reproduce this reasoning, but
you will need to be able to
use both forms of the formula.

How does this work in general? The original formula has a structure that involves dividing one ratio by another. This structure can be expressed mathematically using symbols p, q, r and s as follows.

If $\begin{cases} p = \text{AEI for month } x, \\ q = \text{AEI for same month one year earlier,} \\ r = \text{RPI for month } x, \\ s = \text{RPI for same month one year earlier,} \end{cases}$

then the real earnings are

$$\frac{p}{q} \div \frac{r}{s}.$$

To divide by the fraction $\frac{r}{s}$, you turn $\frac{r}{s}$ upside down and multiply, giving

$$\frac{p}{q} \times \frac{s}{r}.$$

Translating the letter equivalents back into words, this gives an equivalent way of working out the real earnings, compared with one year earlier, as follows.

Real earnings for month x compared with one year earlier $=$

$$\frac{\text{AEI for month } x}{\text{AEI for month } x, \text{ one year earlier}} \times \frac{\text{RPI for month } x, \text{ one year earlier}}{\text{RPI for month } x}.$$

The list of values of the AEI and the RPI in Table 19 will be used to illustrate the use of this formula.

Table 19   Values of the AEI and RPI in 2002 and 2003

	AEI		RPI	
	2002	2003	2002	2003
January	106.3	109.1	173.3	178.4
February	106.9	110.0	173.8	179.3
March	106.7	110.9	174.5	179.9
April	108.0	110.7	175.7	181.2
May	107.9	111.3	176.2	181.5
June	108.2	111.6	176.2	181.3
July	108.4	112.5	175.9	181.3
August	108.6	112.4	176.4	181.6
September	108.8	112.8	177.6	182.5
October	109.0	113.0	177.9	182.6
November	110.0	113.7	178.2	182.7
December	109.5	113.2	178.5	183.5

Source: *www.statistics.gov.uk*

Example 9   *Using real earnings formulas*

Find the increase in real earnings for November 2003 compared with November 2002.

Real earnings for November 2003 compared with one year earlier:

$$\frac{\text{AEI for November 2003}}{\text{AEI for November 2002}} \times \frac{\text{RPI for November 2002}}{\text{RPI for November 2003}}$$

$$= \frac{113.7}{110.0} \times \frac{178.2}{182.7} \simeq 1.008.$$

To express this as a percentage, multiply by 100%, which gives 100.8%.

So real earnings for November 2003 were 100.8% of their value a year earlier.

Activity 30   *Real earnings*

For each of the following months calculate (as a percentage) the real earnings for that month compared with one year earlier.

(a) April 2003   (b) June 2003   (c) September 2003

Table 20 shows the real earnings for each month in 2003 compared with one year earlier.

Table 20   Real earnings for each month of 2003, compared with one year earlier (as a percentage). Calculated from data in Table 19.

Jan.	Feb.	Mar.	Apr.	May	Jun.	Jul.	Aug.	Sep.	Oct.	Nov.	Dec.
99.7	99.7	100.8	99.4	100.1	100.2	100.7	100.5	100.9	101.0	100.8	100.6

For most months in 2003, real earnings compared with one year earlier were greater than 100%. According to this measure, employed people were generally getting slightly better off during the year. However, the percentage increases were all small, and real earnings were slightly less than those of one year earlier in January, February and April.

6.3 So, are people better off?

The central question posed at the start of *Unit 2* was: 'Are people getting better off?' In attempting to answer this question initially, you were asked to consider a simple, everyday measure—the cost of a loaf of bread—and then asked what proportion of a daily wage it represented. However, this measure is clearly inadequate as it looks at only one of thousands of goods and services that people purchase and which affect how well-off they think they are. But the principle of looking at changes in prices compared with changes in earnings was one that was sustained throughout the two units. You have looked at more formal measures, the RPI and the CPI, which monitor changes in prices from large baskets of goods and services, and the AEI, which monitors changes in earnings from a wide range of different occupations and employers.

All the evidence from this section so far suggests that, on average, people in employment *have* been getting better off. For example, Figure 17 indicates that, since 1990, there were only occasional months when the annual rate of inflation was greater than the increase in the AEI over the previous year. For the vast majority of months, the annual increase in average earnings has more than offset the average annual price rise. Of course, this conclusion only applies to the employed; it does not necessarily follow that pensioners, the unemployed or the self-employed have also, on average, been getting better off.

This section ends with a brief look at three wider issues:

◇   the fact that averages are not individuals;

◇   the effect of tax;

◇   whether there is more to being better off than *material well-being*.

Averages are not individuals

It has already been stressed that both the AEI and the RPI are based on *averages* and should not be taken as representing the circumstances of any particular individual or group of individuals. So the AEI and the RPI, even when used together, provide only a very poor assessment of whether a particular group of people is getting better off.

The average being referred to here is a mean and, unlike the median, the mean is often affected by extreme values. It only needs a small group of individuals to secure very large rises in their earnings to produce a substantial increase in the mean earnings. This tells you very little about how everyone else has fared.

It is important to stress, therefore, that the AEI is not sensitive to changes in the overall distribution of earnings—it is only concerned with *averages*, which here means the *mean*.

A further problem with the AEI can be illustrated by the following example.

Example 10 *Mean earnings after redundancy*

In Activity 7 you were introduced to the fictitious firm, Troublefree Computers. The initial earnings of the five employees were £300, £350, £400, £450 and £500. So the mean earnings of the five employees was £400. Suppose that the lowest-paid employee is made redundant. What effect do you think this would have on the AEI?

The mean earnings for the four remaining employees becomes

$$\frac{350 + 400 + 450 + 500}{4} = \text{£}425.$$

The average pay of those remaining employed has gone up!

So, if poorly-paid workers are laid off, then the value of the average pay of those remaining will go up. This means that the AEI will increase! This may seem paradoxical, and is indeed difficult to square with the feeling that if the AEI is increasing then people are generally getting better off.

However, the paradox is somewhat explained when you consider that the AEI is based only on those in employment, so it can increase purely as a result of lower-paid workers being laid off and no longer being counted. On the other hand, it may decrease if unemployed people take new jobs at relatively low pay, even though these people may be better off than when they were out of work.

The effect of tax

A further complication occurs when the effects of income tax are taken into account. A rise in gross earnings of, say, 4% will be unlikely to result in a corresponding rise in net earnings of 4%. There are two reasons for this. First, income is taxed at different rates for different bands of income, so the effect of tax will vary depending on how much is earned. Second, each year the tax allowances tend to be adjusted and this also influences the proportion of earnings that are taxed. So, it is increases in net earnings, and not increases in gross earnings, which must be considered in relation to price changes.

Income tax has, potentially, a role in the redistribution of wealth between rich and poor. In 1979, when there was a change in government from Labour to Conservative, the so-called marginal rate of income tax for high earners (that is, the percentage paid in tax for the last £1 earned) was 83%. By the 1990s, this marginal rate had dropped to 40%.

Another factor to be considered is the balance between direct and indirect taxes. Direct taxes must be paid even if one spends nothing, and include income tax, employees's national insurance contributions and Council Tax. Indirect taxes are essentially taxes on spending, for example, value added tax (VAT). Compared with the higher-paid, the lower-paid need to spend a higher proportion of their earnings. Thus, a larger proportion of their earnings is paid to the government in indirect tax. In 2002–03, taking direct tax and indirect tax together, the households whose income (before tax) was in the top 20% for the country paid on average about 35% of their income in tax, while the 20% with the lowest household incomes paid around 38%.

Source: 'The effects of taxes and benefits on household income, 2002-03' by Caroline Lakin, *Economic Trends 607*, June 2004. The data come from the *Expenditure and Food Survey* that you read about in *Unit 2*.

Is there more to being better off than material well-being?

This section ends with an audio band in which a group of middle-aged and elderly people talked about whether they feel better off. This is an opportunity to broaden the debate from material (and numerical) measures to look at other factors.

Activity 31   *Better off than before?*

Before listening to the audio, note down some other factors which you consider to be important measures of how well-off (or how badly-off) you feel yourself to be.

Now listen to band 5 of CDA5508, called 'Are people getting better off?'. Make a note of some of the factors that were mentioned by those who spoke. The following chart lists all the speakers, in the order in which they speak, and you may wish to make your notes there.

Better off THEN		Better off NOW
	Lil	
	Margaret	
	Anne	
	Rose	
	Connie	
	Randolph	
	Connie	
	Papa	
	Aftab	
	Papa	
	Connie	
	Lil	
	Margaret	
	Rose	
	Lil	
	Rose	
	Lil	
	Aftab	
	Margaret	
	Randolph	
	Lil	

Here are some questions for you to think about after listening.

◇   How might some of these factors be measured?

◇   Did the speakers' factors differ from the items on your list? Why?

◇   Do you feel there is a consensus on what factors are important as measures of how well-off (or how badly-off) people feel?

A clear message of this final subsection has been that making judgements about people's earnings based solely on averages like the AEI can be misleading. Indeed, the same warnings need to be applied to all average measures of earnings and price changes as they say nothing about the inequalities experienced by individuals. Taken overall, average earnings increases do seem to have exceeded price rises during the 1990s and the early years of this century. There seems to be good evidence for claiming that, *if you are still in work*, you are probably better off. However, inequalities in income still persist. So, while the rich have got richer, relatively speaking, the poor are still poor.

Outcomes

After studying this section, you should be able to:

◇   interpret a graph of the AEI or of annual increases in the RPI and the AEI (Activity 28);

◇   use the AEI to calculate the annual increase in average earnings (Activity 29);

◇   use the AEI and the RPI to calculate the real earnings at one date compared with an earlier date (Activity 30);

◇   identify factors other than material ones which affect how well-off people feel (Activity 31).

Unit summary and outcomes

You now know something about how UK government statisticians measure changes in prices and earnings. So you should be able to explain what politicians and journalists really mean when they make broad claims about whether or not 'we' are getting better off. In the course of discussing earnings, several mathematical ideas have been introduced and used—for example, ratios and index numbers, quartiles and deciles, range and interquartile range, boxplots, earnings ratios, and real earnings.

Activity 32 Looking back

Now would be a good time to reflect for a few minutes on your progress over *Units 2* and *3* as a pair. Think about what you knew at the beginning of the units and compare it with what you know now about statistics. Write down what you feel you have gained from studying this unit—for example, a new skill (such as using the calculator), a skill that you have improved or an understanding of some idea or technique. Which topics in this unit have you found straightforward? Which have you found difficult?

Write down an example of something that caused you difficulty and on which you might need to spend more time. If you have identified some aspect of the work in this unit that is causing you real concern, how are you going to go about overcoming this? (For example, you might look back at the preparatory materials.)

There is a printed response sheet for this activity.

Finally, are there any topics that you want to add to your Handbook notes?

Outcomes

You should now be able to:

◇   specify the types of data needed to investigate claims about earnings;

◇   extract relevant information from tables of data;

◇   explain the meanings of the terms 'range', 'lower quartile', 'upper quartile', 'interquartile range', 'lowest decile' and 'highest decile';

◇   explain the meaning of the term 'earnings ratio' and calculate and interpret earnings ratios;

◇   comment on the features of different diagrams (for example, line diagram, boxplot) offered to support your thinking;

◇   find the range, the lower quartile, the upper quartile and the interquartile range for a given batch of data;

◇   draw a rough sketch and an accurate version of a boxplot for a given batch of data with the aid of your calculator;

◇   use your calculator to draw boxplots and calculate various summary statistics (such as quartiles) from batches of data;

◇   draw decile boxplots to represent earnings data given in the form of summary statistics;

◇   make comparisons between distributions using boxplots and decile boxplots;

◇   explain how the RPI (Retail Prices Index) and the AEI (Average Earnings Index) may be used to calculate the 'real earnings', and hence offer a partial resolution to the question 'Are people getting better off?' and discuss the limitations of any conclusions;

◇   identify a range of factors which affect how well-off people feel.

Appendix: Drawing accurate boxplots

Introduction

You have seen how to obtain a boxplot for a batch of data on the screen of your calculator, and how to use this to help you to draw a rough sketch of the boxplot. This appendix provides a procedure for drawing accurate boxplots by hand on squared or graph paper. (It is designed for those who are not confident about tackling Activities 20–22 straight away.)

Drawing an accurate boxplot

This subsection works through the drawing of an accurate boxplot based on the earnings of the eleven male accountants from Table 12 (page 28). The data will be used to illustrate choosing a scale and drawing a boxplot. If you are not happy about drawing boxplots, follow this example through, drawing the boxplot yourself.

Here is a sketch of the boxplot for this batch of data as a guide.

Figure 18   A sketch of a boxplot for the earnings of eleven male accountants

One of the advantages of a boxplot is, unlike a conventional graph, you do not have to read points off it: the summary values are written on it (so are stressed) and the remaining data values are not (so are ignored).

Thinking about a suitable scale is the essential first step in drawing a boxplot. There are several general points to remember that will help you to make a sensible choice.

◇   All the data must fit on the page: the whole range of values in the batch of data, from the lower extreme *min* to the upper extreme *max*.

◇   Choose round numbers as endpoints of your scale; for example, 100 and 500, rather than 107 and 474. This means the lines on the graph paper will mark convenient numbers.

◇   The boxplot must fit on the paper, but you must be able to see clearly all the important features of the batch of data (so not too small).

◇   Make one square on your paper correspond to a convenient whole number such as 2, 5 or 10 (or 20 or 50, etc.), rather than 3 or 7 or 1.8. This will make it easier to find points on the scale.

Choosing the endpoints

Moving out beyond the endpoints to suitable 'round' numbers suggests a scale with endpoints of £300 and £1200, or even £0 and £1200. (£0 and £1200 will be used below.)

Choosing the scale

You now need to work out how to fit the boxplot conveniently and clearly on your paper. The scale you choose will depend on the size of the paper you use but, for this example, suppose that you are using A4 paper. A sheet of A4 paper is approximately 21 cm wide so, as a very rough guide, try to make the useful part of your scale about half to three-quarters of the width of the paper—in the region of 10 cm to 15 cm long. For the male accountants' earnings data, your scale needs to go from £0 to £1200, so that it will cover a range of £1200. If you make this correspond to 12 cm, then £100 will correspond to 1 cm.

Drawing the scale

You are now in a position to draw the line on which the scale will be marked—called the *axis*. First, decide where to put your line. Leave some space above the line in which to draw the boxplot: 3 cm should be sufficient. Now draw a horizontal line across the page, longer than the length you need to use (in this case, longer than 12 cm). Larger values will be towards the right-hand end of the axis, indicated by an arrowhead as shown below. To show the scale, mark the axis with vertical lines at suitable regular intervals; in this case, every 2 cm or £200. For the male accountants' earnings data, the range of £1200 is to correspond to 12 cm, so if you mark points at 2 cm intervals from 0 to 1200 then you will have seven marked points as shown in Figure 19.

Figure 19   Draw the axis

Label the axis with numbers, and write the quantity measured (earnings) and the units (£) in which the values are measured below the arrow, as shown in Figure 20. Although £100 corresponds to 1 cm, you do not need to mark every point. You could mark intervals of £200 or even £500.

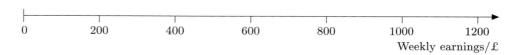

Figure 20   Label the axis

At this point, pause to ask yourself the following questions. Is the scale easy to use? Will the intervals make for easy plotting? That is, can you find points easily on the scale? (For example, can you find the point corresponding to £480 easily?) If the scale is unsatisfactory, go back and choose a more suitable scale.

The box and whiskers

Having decided on and drawn the scale, you are now ready to draw the box and whiskers and label the key points on the boxplot as in your rough sketch.

First, draw the box about 1 cm above your axis. Nothing is represented by the *thickness* of the box. Your diagram will look better if the box is neither very thin nor very fat. Draw vertical lines in line with the positions on the scale of the lower and upper quartiles (536 and 881 in this example). Complete the box by joining the two vertical lines. Mark the median (649 in this instance) by drawing a vertical line through the box in line with the position of the median on the scale, as shown in Figure 21.

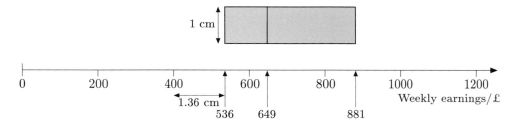

Figure 21   Draw the box

Mark the two extremes, *min* and *max*, by two short vertical lines in line with the positions of *min* and *max* on the scale (376 and 1117 in this instance). Position them so that each can be joined halfway up the box, as shown in Figure 22.

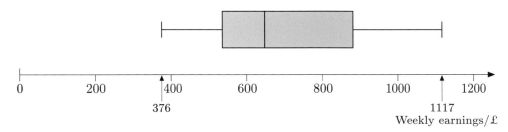

Figure 22   Draw the whiskers

Complete the boxplot, by transferring the five values from the rough sketch, as shown in Figure 23.

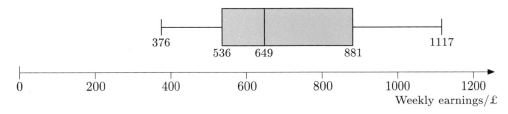

Figure 23   Label the key values

When you have drawn your boxplot, you might like to compare it with the boxplot drawn on the screen of your calculator to check that they are the same shape and that you have not made any errors.

You can also draw two or more boxplots using the same scale, as a useful way of comparing two batches of data. If you are drawing two boxplots on the same scale, then it may be easier to draw one above the scale and the other below the scale. But you may find it easier to compare them if they are both above the scale, next to each other. Make sure you draw the boxes of both boxplots the same width, so that your diagram does not give a misleading impression. Label both boxplots clearly to distinguish between them. This is shown in Figure 24 for the male and female accountants' earnings (using the data from Table 12).

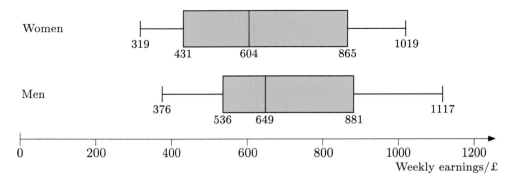

Figure 24   Boxplots of gross weekly earnings of accountants

Now return to Section 4 and do Activities 20 to 22 (starting on page 38) to practise what you have just learned.

Comments on Activities

Activity 1

See the comments after the activity.

Activity 2

You may have chosen the absolute difference or the relative difference.

See the comments after the activity.

Activity 3

(a) Women worked 37.4 hours per week on average—fewer hours than the average worked by men, which was 40.9 hours.

(b) On average, men did 1.5 hours more overtime per week than women $(2.2 - 0.7 = 1.5)$. Alternatively, men did over three times as much overtime as women $(2.2/0.7 \simeq 3.14)$.

(c) Removing overtime pay from the gross weekly earnings figures would reduce the men's figure more than the women's figure, since men did more overtime. You would expect this to narrow the 'gap' between men's earnings and women's earnings and therefore *increase* the earnings ratio.

(d) Since men worked more hours per week on average than women, you would expect men's gross weekly earnings to be more than women's, even if they were paid the same for similar amounts of work. A fairer comparison might be to look at the gross *hourly* earnings of men and women. This would eliminate the effect on earnings of men working more hours per week than women.

Activity 4

(a) The earnings ratio at the mean based on weekly earnings excluding overtime is

$$\frac{389}{500} = 0.778 \text{ or approximately } 78\%.$$

(b) The earnings ratio at the mean based on hourly earnings is

$$\frac{1056}{1288} \simeq 0.8199 \text{ or approximately } 82\%.$$

(c) See the comments after the activity.

Activity 5

(a) The earnings ratio at the mean for each year is given in the table below.

Year	1989	1991	1993	1995	1996	1997
Earnings ratio (%)	76	78	79	79	80	80

Year	1998	1999	2000	2001	2002	2003
Earnings ratio (%)	80	81	81	82	81	82

(b) There was a slow, steady increase in the earnings ratio over this period, up to 2001 at least.

(c) Since the earnings ratio has increased, this measure suggests gender inequalities in earnings have narrowed.

Activity 6

(a) The earnings ratios at the median are given in the table below.

Earnings ratios at the median	(%)
Gross weekly earnings including overtime	78
Gross weekly earnings excluding overtime	83
Gross hourly earnings excluding overtime	87

(b) As was the case when using the mean, the earnings ratio at the median increases when overtime is excluded and again when hourly earnings are considered instead of weekly earnings. In each case, the earnings ratio at the median is higher than the corresponding earnings ratio at the mean.

Activity 7

(a) The mean earnings and the median earnings required in parts (a) to (c) are given in the table below, together with the mean and median for several other possible values of the manager's earnings.

Manager's earnings	Mean	Median
500	400	400
600	420	400
700	440	400
800	460	400
900	480	400
1000	500	400
5000	1300	400

(b) See part (a).

(c) See part (a). The median is unaffected by the increases in the manager's earnings, whereas the more the manager's earnings increase, the greater the mean becomes.

Activity 8

(a) 50% of people earn less than the median wage, so a person of median earnings will pass by halfway through the parade at 10.30 am.

(b) You were told that a person of mean height passes by twenty minutes before the end of the parade—that is, at 10.40 am. Since 10.40 am is 40/60 or $\frac{2}{3}$ of the way through the hour, $\frac{2}{3}$ or 67% of people earn less than the mean wage.

(c) Some images can be misleading!

In the cartoon, a person earning twice the average wage is drawn twice as tall as a person earning the average wage. However, that person is also drawn twice as wide—the tall people are not tall and thin—so the *area* of the cartoon taken up by a person earning twice the average wage is *four* (2×2) times the area taken up by a person earning the average wage. And, in practice, a reader may well interpret a person in the cartoon as a figure occupying a *volume* in space. So the impression received is of a figure *eight* ($2 \times 2 \times 2$) times as large. Thus, the effect of the cartoon is to exaggerate the differences in earnings of different people.

Unfortunately, many published diagrams make use of area or volume to exaggerate the visual effect of points they are trying to make: look out for this whenever you see diagrams used to support an argument.

In addition, the cartoon is based on the idea that greater height corresponds to greater income. This choice has quite strong psychological overtones to do with cultural norms of 'stature', 'importance', and so on: it is a far from neutral image. Consider the impact of a redrawn cartoon where the key image was a person with their hand outstretched: the larger the salary, the longer the arm.

One reason you may be able to orientate yourself with regard to the cartoon as it stands is that you have plenty of experience of the distribution of people's heights to bring to bear on interpreting this image. Most importantly, there is no scale, other than the notion of 'average height'.

Activity 9

To investigate whether or not women are being paid less than men for similar work, data are needed on the earnings of men and women in individual occupations. Ideally, these data would exclude overtime and, since men tend to work longer hours than women, hourly earnings would be desirable. To find out whether women are employed predominantly in occupations with low pay, you need to know how many men and women are employed in the different occupational groups.

Activity 10

The percentages of women and men in various occupational groups are given in the table below.

Major Occupational Group	Women	Men
Managers and Senior Officials	12	19
Professional Occupations	14	13
Associate Professional and Technical Occupations	16	14
Administrative and Secretarial Occupations	31	8
Skilled Trades Occupations	2	15
Personal Service Occupations	8	2
Sales and Customer Service Occupations	7	4
Process, Plant and Machine Operatives	3	14
Elementary Occupations	6	13
Total	100	100

The percentages in some of the groups are very different for women and men. There is a much greater proportion of women than of men in 'Administrative and Secretarial Occupations' and 'Personal Service Occupations', and to a lesser extent in 'Sales and Customer Service Occupations'. There are much greater proportions of men than of women in 'Skilled Trades Occupations', 'Process, Plant and Machine Operatives' and 'Elementary Occupations', and to a lesser extent in

'Managers and Senior Officials'. So the patterns of employment of men and women are very different.

One effect of using a table is to allow the comparisons between men and women to be made more directly: the eye moves easily from left to right and up and down. Also, the percentage totals can be used easily to check the arithmetic. (In fact, the columns do not quite add up to exactly 100% because of rounding.)

Activity 11

The earnings ratio at the median for chefs and cooks is:

$$\frac{600}{670} \simeq 0.8955 \text{ or approximately } 90\%.$$

The earnings ratio at the median for cleaners and domestics is:

$$\frac{544}{569} \simeq 0.9561 \text{ or approximately } 96\%.$$

Using hourly earnings instead of weekly earnings has increased the earnings ratio in each case.

Activity 12

One possible explanation is that proportionately more women than men in each occupational group are in junior, lower-paid posts and proportionately fewer women than men are in senior, more highly-paid positions.

Activity 13

For the women's batch,

$$\text{range} = max - min = 1019 - 319 = 700.$$

For the men's batch,

$$\text{range} = max - min = 1117 - 376 = 741.$$

The range of the earnings is greater for the men's batch than for the women's batch.

Activity 14

The main weakness of all three measures is that they do not take any account of the size of a batch of data. Consider proposed measure (b), for instance. For a small batch, excluding four values at each end may involve excluding all, or nearly all, of the values in the batch. For a large batch, excluding only four values may not exclude all the very unusual values. So the measure still may not accurately represent the spread of the main body of the data. Similarly, if only one value is excluded, as with measure (a), then an unusually large or small value may still remain. And using measure (c) for a small batch may mean that all the values are included, whereas using it for a large batch nearly all the values are excluded.

A balance is needed between excluding so much data that the measure is not representative of the batch of data and excluding so few values that the results are still strongly influenced by a few extreme values.

Activity 15

(a) For a batch of nine values, the configuration looks like the figure below.

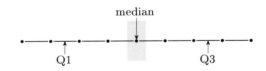

(b) A batch of ten values produces the following diagram.

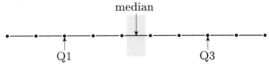

Activity 16

(a) The diagram for eleven values looks like the following.

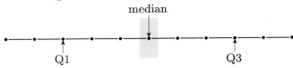

(b) Men's earnings

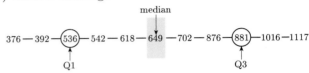

The diagram above acts as a data template and shows which are the key data values. Matching the numerical data with the image, gives the following values:

$$\text{median} = £649,$$
$$Q1 = £536,$$
$$Q3 = £881.$$

(c) For the batch of women's earnings:

$$\text{median} = £604,$$
$$Q1 = £431,$$
$$Q3 = £865.$$

Activity 17

The interquartile range for the batch of women's earnings is

$$Q3 - Q1 = £865 - £431 = £434.$$

This is rather larger than the interquartile range for the batch of men's earnings (£345). Using this measure, the earnings of the female accountants are more widely spread than the earnings of the male accountants.

Activity 18

(a) The quartiles and extremes are given below.

	min	*Q1*	*Q3*	*max*
Women	495	519.5	678	1010
Men	531	622.5	890.5	981

Female chefs and cooks:

range $= 1010p - 495p = 515p$;

interquartile range $= 678p - 519.5p = 158.5p$

Male chefs and cooks:

range $= 981p - 531p = 450p$;

interquartile range $= 890.5p - 622.5p = 268p.$

(b) The range of earnings is greater for the women than for the men. The interquartile range is however greater for the men than for the women, indicating that the greater spread for female chefs and cooks, suggested by the range, may be solely due to one woman earning an unusually large or small amount. (In fact, the best-paid woman in the batch does earn much more than any of the others.)

Activity 19

(a) The earnings of female chefs and cooks are generally lower than the earnings of male chefs and cooks. For example, the upper quartile for the women is somewhat below the median for the men, so over 50% of the men earned more than 75% of the women. Four out of the five values which are marked on the boxplot are lower for the women than the corresponding values for the men. The box is longer for the men than for the women. So there is a greater spread in the earnings of the men than in the earnings of the women (as measured by the interquartile range). However, the right whisker is much longer for the women than for the men, so the women have a greater spread of earnings than the men as measured by the range. (But the length of the right whisker and the range for the women are very much influenced by one woman, the highest

earner, who earned far more, 1010p, than the second highest, 689p.)

(b) In both cases, the whiskers and the box indicate slightly different things about skewness. For the women, the box is almost symmetrical, with both halves of roughly equal length. But the right-hand whisker is much longer than the left, indicating that the data are right-skewed. For the men, the box strongly indicates right skew, but the whiskers are roughly equal. Overall, both earnings distributions appear right-skewed on the whole.

However, you should bear in mind that all the comparisons in parts (a) and (b) are based on rather small batches of data—it would not be wise to use these to draw definite conclusions about the earnings of chefs and cooks in general.

(c) Two boxplots of related batches of data drawn to the same scale make for very direct comparison: all you need to do is look up and down at any point. Because boxplots have a uniform structure in what they stress (quartiles, median, extremes) and what they ignore (most actual data values), looking at the key features makes a simple comparison of two batches easy.

Activity 20

Moving out beyond the endpoints to suitable 'round' numbers suggests a scale with endpoints of 400p and 1100p (say), so the scale must cover a range of 700p. The actual scale will depend on the amount of space you have available. The boxplots for chefs' and cooks' earnings are shown below.

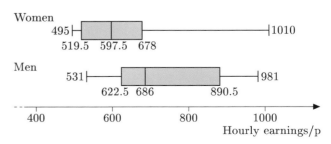

Boxplots of earnings data for chefs and cooks

69

Activity 21

(a) The sketches should look (roughly!) like the accurate diagram in part (b), but without the scale.

(b) In this case, possible 'round' numbers beyond the endpoints are £200 and £800, a range of £600. Taking £50 to correspond to 1 cm produces boxplots which fit comfortably on an A4 page. Boxplots for the nurses' earnings are shown below.

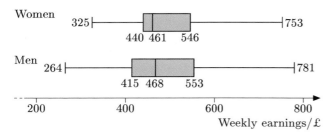

Boxplots of earnings data for nurses

(c) There is no marked difference in one direction or the other between the women and the men. Earnings for the men are rather more spread out.

Activity 22

(a) The sketches should look like the accurate diagram in part (b), but without the scale.

(b) Boxplots for the earnings of the secondary education teachers are shown below.

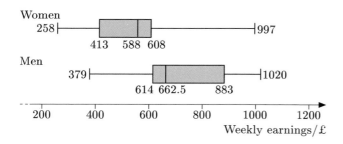

Boxplots of earnings data for secondary education teachers

(c) The earnings of the men were generally higher than those of the women. For example, the upper quartile of the women's earnings is less than the lower quartile of the men's earnings. All five values on the boxplots are lower for the women than for the men. The box is longer for the men than for the women, indicating a larger interquartile range for the men. However the range is larger for the women, though, as in Activity 19; this is because of a single high-earning woman.

Activity 23

(a) 50% of the men earned between £548 (the lower quartile) and £817 (the upper quartile).

(b) 25% of the women earned £461 or less, so £461 is the largest amount earned by the lowest-paid 25% of women.

(c) Approximately 25% of the men earned more than £817; and approximately 75% of the women earned less than £718.

Activity 24

(a) The highest decile for the female accountants is £874, so 10% of the group earned £874 or more.

(b) £374 is the lowest decile, so 10% earned £374 or less.

(c) £546 is the median earnings and £874 is the highest decile. 40% of the women (50% − 10%: see Figure 14) received more than the median earnings but less than the highest decile, so 40% of the women earned between £546 and £874.

(d) 25% − 10% = 15% of the women earned between £374 (the lowest decile) and £461 (the lower quartile).

Activity 25

All five key values shown are higher for men than for women, so generally male accountants from this survey earn more than their female counterparts. In particular, the median for women is similar to the lower quartile for men, showing that the lowest paid 50% of female accountants all earn less than 75% of the males. The spread of earnings is greater for men than for women within both the box and the whiskers.

Both earnings distributions are clearly right-skew.

Activity 26

The decile boxplots below represent the gross hourly earnings of chefs and cooks.

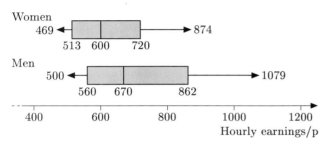

Men earn more than women at all five of the key values shown. So, on average, male cooks and chefs earn more per hour than female cooks and chefs. However, differences were greater at the higher end of the earnings distributions. The spread of the earnings was greater for the men than for the women, both within the box and in the whiskers. Both earnings distributions are right-skew.

Activity 27

The earnings ratios are given in Table 21 below.

Table 21 Earnings ratios for chefs and cooks

Highest decile	81%
Upper quartile	84%
Median	90%
Lower quartile	92%
Lowest decile	94%

These figures show that at all levels, male chefs and cooks earn more than female chefs and cooks. In the lower half of the earnings range, female chefs and cooks do not receive much less pay from men for an hour's work: the earnings ratios at the lowest decile, lower quartile and median are 94%, 92% and 90% respectively.

However, higher-paid women fare less well: the earnings ratios decrease steadily from the bottom of the distribution to the top. At the upper quartile, the earnings ratio is 84% and at the highest decile it is only 81%.

Activity 28

There are a number of features in the graph that you may have commented on. Perhaps the clearest features are the peaks that occur around December 2000, April 2002, November 2002, March 2003 and particularly January 2004. These cannot relate to regular, yearly, pay changes around Christmas because the seasonal adjustment of the index would allow for that. However, many workers get annual bonuses, often near the turn of the year. In some businesses, bonuses are large but can vary considerably from year to year. Because of this variability, seasonal adjustment does not remove the effect of these bonuses, therefore these peaks probably have something to do with bonuses. (In fact, a version of the AEI that excludes bonus payments is also published, and a graph of that version does not show these peaks.) You may well have spotted other features and suggested other explanations.

Activity 29

The value of the AEI in June 2004 as a percentage of its value in June 2003 was

$$\frac{116.3}{111.6} \times 100\% \simeq 104.2.\%$$

So the AEI increased by 4.2% of its June 2003 value between June 2003 and June 2004.

Activity 30

(a) The real earnings for April 2003 compared with one year earlier are:

$$\frac{\text{AEI, Apr. 2003}}{\text{AEI, Apr. 2002}} \times \frac{\text{RPI, Apr. 2002}}{\text{RPI, Apr. 2003}} \times 100\%$$

$$= \frac{110.7}{108.0} \times \frac{175.7}{181.2} \times 100\%$$

$\simeq 99.4\%$. In this case, real earnings *fell* slightly over the year to April 2003.

(b) The real earnings for June 2003 compared with one year earlier are:

$$\frac{\text{AEI, June 2003}}{\text{AEI, June 2002}} \times \frac{\text{RPI, June 2002}}{\text{RPI, June 2003}} \times 100\%$$

$$= \frac{111.6}{108.2} \times \frac{176.2}{181.3} \times 100\%$$

$$\simeq 100.2\%.$$

(c) The real earnings for September 2003 compared with one year earlier are:

$$\frac{\text{AEI, Sept. 2003}}{\text{AEI, Sept. 2002}} \times \frac{\text{RPI, Sept. 2002}}{\text{RPI, Sept. 2003}} \times 100\%$$

$$= \frac{112.8}{108.8} \times \frac{177.6}{182.5} \times 100\%$$

$$\simeq 100.9\%.$$

Activity 32

Everybody's answer will be different. You may well have mentioned using ratios again, and learning to draw boxplots with or without the calculator, as skills you have developed. You may have found the discussion about the formula for real earnings difficult. Did using letters instead of long strings of words help or not? If you still have major concerns or difficulties, you should consider contacting your tutor about them.

Activity 31

An example of the completed chart for this activity is as follows. Yours will differ!

Better off THEN		Better off NOW
• free from pressures	Lil	
• companionship and simple pleasures	Margaret	
• safety at night; money went a long way	Anne	
	Rose	• more poverty now; range of home activities
	Connie	• racism in housing is less now
• homelessness now	Randolph	
	Connie	• racism in housing is less now
• cost of housing was less then	Papa	
• cost of living was less then	Aftab	
• cost of living was less then	Papa	
	Connie	• pensions are better now, particularly for women
	Lil	• horrors of housework in the past
	Margaret	• laundry and general housework harder in the past
	Rose	• fashions for older people now
	Lil	• older people can use the hairdressers now
	Rose	• hair care was primitive then
	Lil	• hair care unaffordable then
• public services have deteriorated	Aftab	
	Margaret	• lack of money then, no pressure of work now, more home ownership now
• computers have brought unemployment	Randolph	
	Lil	• has regular pension, so no employment worries

Acknowledgements

Grateful acknowledgement is made to the following sources for permission to reproduce material in this unit:

Illustration

p. 19: Low Pay Unit.

Cover

Guillemots: RSPB Photo Library; Sellafield newspaper headline: *Independent*, 8.1.1993; other photographs: Mike Levers, Photographic Department, The Open University.

Index